Martial Rose Library
Tel: 01962 827306

To be returned on or before the

THE ECONOMIC GEOGRAPHIES OF ORGANIZED CRIME

TIM HALL

THE GUILFORD PRESS
New York   London

Copyright © 2018 The Guilford Press
A Division of Guilford Publications, Inc.
370 Seventh Avenue, Suite 1200, New York, NY 10001
www.guilford.com

All rights reserved

No part of this book may be reproduced, translated, stored in a
retrieval system, or transmitted, in any form or by any means,
electronic, mechanical, photocopying, microfilming, recording,
or otherwise, without written permission from the publisher.

Printed in the United States of America

This book is printed on acid-free paper.

Last digit is print number: 9 8 7 6 5 4 3 2 1

Library of Congress Cataloging-in-Publication Data

Names: Hall, Tim, 1968– author.
Title: The economic geographies of organized crime / Tim Hall.
Description: New York : Guilford Press, [2018] | Includes bibliographical
  references and index.
Identifiers: LCCN 2017039047| ISBN 9781462535194 (pbk.) | ISBN
9781462535200 (hardcover)
Subjects: LCSH: Organized crime. | Economic geography.
Classification: LCC HV6441 .H35 2018 | DDC 364.106—dc23
LC record available at https://lccn.loc.gov/2017039047

For Cath

PREFACE

The origins of this book lie in a British national newspaper article (now long lost) that I read sometime during the summer of 2006. The article made the claim that somewhere between 15 and 20% of the world economy was linked to organized crime. That the figure was this high astonished me. At the time, I was teaching an undergraduate course in economic geography (at the University of Gloucestershire). I was not an economic geographer and was assigned the course when I started at the university because no one else wanted to teach it and I was the most junior member of the geography department. Over 10 years later, I had still not managed to offload the course onto anyone else. On reading the newspaper article, my first thought was that I could drop a session on the economic geographies of organized crime into the course. It seemed to be an opportunity to add some excitement to what was a somewhat dry class and I could see it raising all sorts of questions about the risks and difficulties of researching illicit and illegal economic activities, about the spatialities and networks of industries like drug production and maritime piracy, and about the data sources available to examine clandestine economic markets. Duly inspired, I wrote the session title into my lecture schedule and thought no more about it for some weeks.

About 2 weeks before I was due to deliver the session, I sat at my desk with the pile of economic geography textbooks upon which I had based my teaching on the subject. I flicked through the first book looking for the chapters on illegal economic activities, illicit economic activities, organized crime, and criminal markets. Nothing. I turned to the index. Nothing. I picked up a second book and again found no reference to any relevant material in that book. So it went on. By the end of the pile, I was panicking. All I found was a single diagram in one text (Hudson, 2005: 6, reproduced here as Figure 1.1) that acknowledged the presence of illicit and illegal economic activities in contemporary capitalism. Even here, though, there was no discussion of these economic realms in the surrounding text. This absence of any discussion of organized crime specifically, and illicit and illegal economic activities more generally, felt like something of a black hole within economic geography and one that, unless I managed to find some material from somewhere before I was due to deliver the session to my students, I was likely to fall into. Luckily for me, our university library had a small criminology section and I was able to pull enough material together to deliver a vaguely coherent set of ideas to my students.

To my surprise, the session was one of the most successful I had ever delivered. The students in the class found the material and ideas fascinating. They all had knowledge of organized crime, often derived from films such as *The Godfather*, *Goodfellas*, or *Lock, Stock and Two Smoking Barrels* and video games such as *Grand Theft Auto*, and, in the case of one Polish Erasmus exchange student, who had contributed nothing to classroom discussions in any of the previous sessions, from living in a region where illicit economic activities were part of his everyday experience. At the end of the course two students (Emily Huntley and Martin Robson) chose to look at organized crime for their main assessments. They produced beautifully geographical accounts of money laundering (Emily) and drug production and trafficking (Martin), both of which received first-class marks. They spoke of the concentration of these activities in certain parts of the world, the reasons for this concentration, the transnational connections that moved illicit finance and drugs around the world, and the impacts of these industries on the spaces they touched. On reading their work, I was struck by the thought that if my own second-year undergraduate students could speak of these major (illegal) industries in deeply geographical ways, then why had no economic geographers, who were widely acknowledged as being at the forefront of critical accounts of the global economy, done likewise? In not recognizing and discussing the extensive illegal markets that characterized the global economy, they were, it seemed to me, producing inherently partial accounts.

I formulated a plan to write a short comment article for a journal arguing little more than that economic geographers should acknowledge more fully illicit and illegal markets if they wanted to produce more complete discussions of the contemporary global economy. I recognized an empirical gap here and felt that this perhaps opened up some conceptual shortcomings. I ended up extending this side project and producing a series of articles and book chapters that explored what economic geographies of organized crime might look like (Hall, 2010a, 2010b, 2012, 2013). The reception that this work received from editors, referees, readers, and students was in marked contrast to the lukewarm receptions that much of my previous research into urban geography had received. I was convinced I was on to something, and since the late 2000s these efforts have come to dominate my research activities and increasingly my research output.

Having delved into these questions in more detail since 2006, I have discovered something of a tradition, albeit a very limited one, of geographical inquiry into aspects of organized crime as well as an interest among criminologists in many of the spatial aspects of organized criminal activity (these are discussed in the early chapters of this book). However, these traditions are relatively underdeveloped conceptually and empirically. In the last few years, geographical interest in the illicit has undoubtedly shown signs of growth. Geographers such as Andrew Brooks, Ray Hudson, and Craig Martin are pursuing various interesting lines of inquiry that reveal the economic, cultural, and political geographies of the illicit. Increasingly geography conference sessions are also bringing together geographers and others from cognate disciplines working on questions about the contemporary and historical spaces of the illicit and illegal. These perspectives are increasingly critical and have established that illicit and illegal economic practices are not purely the preserve of gangsters. More and more they seem to be embedded in the corporate and political mainstream.

This book is an attempt to interrogate and bring together previously discrete bodies of literature and in doing so to try to develop more robust conceptual foundations for geographically informed discussions of organized crime and its markets. There is much that remains to be done, though, particularly in terms of empirical investigation of the economic geographies of organized crime. As we move into an apparently evermore unequal, unstable, and opaque economic age, these questions seem even more relevant and urgent than they did when I first began to consider them in 2006.

ACKNOWLEDGMENTS

This book is the product of many places and the people I have encountered therein. Without doubt, I would not have spent the last 10 years looking at the economic geographies of organized crime without the enthusiasm of my students and the ideas that they have articulated in class and through their assessed work. These include students in the courses "Living in a Global World," "Economic Change and Location," and "Global Crime" (University of Gloucestershire) and "Global Risks" and "Globalized and Organized Crime" (University of Winchester), as well as the colleagues who have co-taught those classes with me over the years, including Mick Healey, Jon Hobson, Kenny Lynch, Charlie Parker, and Dave Turner (Gloucestershire) and Tom Ball, Matt Clement, and Vincenzo Scalia (Winchester).

The Winchester undergraduate students Nykole Webb and Callum Whitehouse did some remarkable research assistance work that contributed hugely to my understanding of global measures of illicit and illegal economic activities. Their work was conducted under the University of Winchester Research Apprenticeship Programme, an innovative program that pairs undergraduate students with researchers to work together on research projects. All universities should have schemes such as this that promote collaborative working relationships between students and staff.

Many of the ideas in this book have been aired and discussed at various conferences and workshops over the last few years. These include the annual conferences of the British Society of Criminology (Huddersfield, 2008), the Royal Geographical Society (with the Institute of British Geographers) (Manchester, 2009; London, 2010), the Geographical Association (Guildford, 2011), the Association of American Geographers (Los Angeles, 2013; San Francisco, 2016), and the Hampshire Geographical Association (Winchester, 2014); the convention of the International Association of Political Science Students (Prague, 2015); the workshops "Turbulent Trade Routes" (Lancaster, 2010) and "The Space of the Illicit in the City" (Venice, 2016); and my inaugural lecture "The Problems of Organized Crime" at the University of Winchester (2015). Many thanks to the organizers of these conferences, lectures, and workshops for the invitations to participate; to my fellow presenters; and to the audiences for their contributions.

The Regional Studies Association provided funding for a research network I co-led entitled "Illicit Actors, Regional Governance and Development" (2014–2016). Many thanks to the Regional Studies Association for this funding and to Ray Hudson and Francesco Chiodelli for their roles in the development and organization of the network and to the contributors to the three events held as part of this research network (Winchester, 2015; London, 2016; and L'Aquilla, 2016). Those discussions, running in parallel to the writing of this volume, undoubtedly informed its development.

The Guilford Press has been a very supportive and accommodating publisher, particularly with regard to the lengthy writing time I requested due to other professional commitments. My thanks go to Seymour Weingarten for initially approaching me with the idea for the book and to C. Deborah Laughton and Katherine Sommer for all of their help and support during the book's production.

Friends and family have been very important to me during the writing of this book. Particular thanks are due to my wife, Cath, for the copyediting and for generally keeping me sane; to my brother-in-law and sister-in-law, Joss and Janie White; and to my nephew, Jake White, for all the conversations about the Mafia and Somali piracy. Any shortcomings of the book are entirely Jake's fault and have nothing to do with me.

CONTENTS

Chapter 1   Geography and Organized Crime   1
*Definitions and Discourses
    of Organized Crime   3
Economic Geography Perspectives   9
The Political and Social Production
    of Organized Crime   13
Prior Geographical Literatures
    of Organized Crime   16
The Book   18*

Chapter 2   Contemporary Organized Crime   21
*The Economies of Illegal Drugs   23
Other Commodity Trafficking   28
People Trafficking   30
Cybercrime   31
Environmental Crime   33
Money Laundering   34
Organized Crime and the State   35
Criminal Organization
    in a Global Economy   39*

Chapter 3   Measuring and Researching
Organized Crime   41
*Official Measures of Organized Crime   43
Perception and Experience
    Surveys of Organized Crime   49*

    Organized Crime Indexes 50
    Threat and Risk Assessments 53
    Investigative Journalism Sources 53
    True Crime and Gangster Autobiographies 56
    Social Media and Online Sources 58
    Historical and Ethnographic
     Accounts of Organized Crime 60
    Critical Geopolitics 65
    Policymaking and Knowledges
     of Organized Crime 66

Chapter 4 The Organization of Criminal Enterprises 71
    Forms of Criminal Organization 73
    Criminal Organization
     under Post-Fordism 77
    Regional Differences in the Organization
     of Criminal Groups 88
    Regulating Criminal Economies 92

Chapter 5 The Spatialities of Organized Crime 98
    Global Economic Contexts 100
    Local Geographical Contexts 110
    Multiscalar Approaches 117
    Network Ontologies 119

Chapter 6 Criminal Mobilities 123
    Networked Criminal Mobilities 126
    Criminal Commodity Movements
     and Their Spaces 130
    Representations of Illicit
     Commodity Movements 137

Chapter 7 Responding to Organized Crime 152
    Extant Responses to
     Organized Crime 156
    Changing the Policy Environment 169
    Geographical Issues
     and Policy Responses 175

	Contents	
Chapter 8	Toward Economic Geographies of Organized Crime *Retrospect 182* *Prospect 185*	181
	References	189
	Index	213
	About the Author	223

CHAPTER 1

GEOGRAPHY AND ORGANIZED CRIME

INTRODUCTION

Globally the economy has been conceptualized as four overlapping, intersecting realms. These are formal capitalist economies, various noncapitalist economies, informal economies, and illegal capitalist economies (see Figure 1.1). Economic geography has been at the forefront of attempts to produce sophisticated, spatialized readings of these realms. However, these discourses have advanced only partial interpretations that have been overwhelmingly rooted in licit economic realms and where they have engaged with illicit and illegal economic activities, their treatment has been highly uneven. While there has been an extensive literature that has explored the geographies of informal economies and their spaces (Herwartz, Tafenau, and Schneider, 2015; Leyshon, Lee, and Williams, 2003; Samers, 2005; Williams, 2004, 2006), there has been little from within economic geography, or any of the other subdisciplines of human geography, that has examined illegal and illicit economies, such as those associated with organized criminal activity, to any substantive degree (Boyce, Banister, and Slack, 2015: 448; Hall, 2010a: 841, 2013; Hudson, 2014: 776). Taylor, Jasparro, and Mattson (2013), for example, note this specifically with regard to geographical work on illegal drugs.

This reluctance to engage with illegal economies is not unique to economic geography. Others have noted lacunae across social science

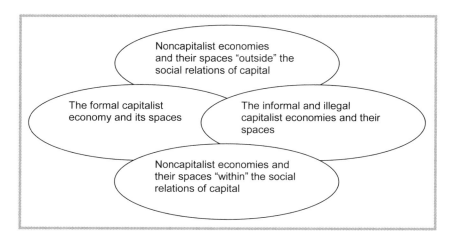

FIGURE 1.1. Relations between economies and their spaces in capitalism. From Hudson (2005: 6). Reprinted by permission of Sage.

literatures when it comes to issues such as economies that are dominated by entrepreneurial criminal groups, the transaction of illegal and illicit commodities, and the movements and networks associated with them (Brown and Cloke, 2007; Castells, 2000: 171; Hudson, 2014: 776; Kirby and Penna, 2011: 184). There are obvious barriers to researchers wishing to engage with these economies and their actors, such as safety, access to these secretive worlds, and the reliability of data (Hobbs and Antonopoulos, 2014; Holmes, 2016). These are issues that are picked up in more detail in Chapter 3 of this book. But, despite these barriers, equally there are a number of imperatives that suggest taking these realms more seriously is a timely and important endeavor. This book will consider many of these issues.

This book is written with two audiences in mind who come from very different disciplinary and subject backgrounds and for whom it is difficult to plot a common narrative journey. It is written for students and scholars of economic geography for whom the terrain of investigation may have been traditionally and predominantly sketched out by the contours of the formal, licit economy. This book aims to highlight the ways in which illicit and illegal markets, services, and commodities and the groups operative within these markets are also as subject to geographical interpretation as any from within the licit economy. The insights of these readers are particularly invited as there are potentially many innovations from the conceptual, methodological, and epistemic work of economic geography that have simply not been applied to the study of the illicit and illegal economies of organized crime to any great extent to date. While this book attempts to begin this endeavor for one of the first times in any systematic way, there remains a potentially vast amount of work to do within the context of this broader project. These readers, though, will be familiar with some of the more basic articulations of economic geography perspectives that appear at various points throughout this book. They might profitably focus their reading of this text on those sections that discuss organized crime rather than economic geography.

The other audience of this book is that diverse collection of students and scholars of organized crime who approach it from a number of different disciplinary perspectives such as anthropology, criminology, political science, and sociology. Within these literatures there has, so far, been only a patchy, and often implicit, engagement with the spatial dimensions of the issue (Hall, 2012, 2013). It is likely that this audience, while familiar with certain literatures on organized crime, will be seeking new insights that the spatially informed perspective within this book can offer. For these readers, those sections that outline the potential contributions of economic geography to the understanding of organized crime should be particularly useful. Whereas the discussions of examples of organized criminal groups and markets might offer empirically familiar terrains in some cases for these readers, they are encouraged to engage with these sections and to critically reflect upon what their articulation through an economic geography lens adds to the accounts that have emerged from within their own disciplines.

DEFINITIONS AND DISCOURSES OF ORGANIZED CRIME

The concept "organized crime" frequently evokes powerful cultural imagery, typically rendered through exclusive foci on a narrow range of "rock star trades" (Nordstrom, 2011: 12), which is often at odds with its more parochial yet complex empirical realities (Hobbs, 1998; Levi, 2002, 2014; Woodiwiss, 2003; Woodiwiss and Hobbs, 2009). This is almost as true of definitions of organized crime as it is of its more unreliable cultural discourses (Chibnall, 2009; Larke, 2003; Rafter, 2006). Organized crime, then, has long been recognized as a contested concept and one that eludes easy or satisfactory definition (Levi, 2002, 2014; Madsen, 2009; Maltz, 1976; Wright, 2006). Often definitions both from the criminological literature and official sources have sought to isolate "essential" (Levi, 2002: 880) structural or behavioral characteristics of organized crime. Albanese argues, for example, that organized crime is "a continuing criminal enterprise that rationally works to profit from illicit activities that are often in great public demand. Its continuing existence is maintained through the use of force, threats, monopoly control, and/or the corruption of public officials" (2005: 10; for other discussions of definitions of organized crime, see Calderoni, 2011; Levi, 2002; Madsen, 2009; Wright, 2006: 2–14).

The types of criminal activities that are routinely grouped together under the banner of "organized crime" include the trafficking of many types of commodities including narcotics, weapons, people, various forms of stolen and counterfeited goods, nuclear materials, precious metals, stones and other natural resources—which might originate from within conflict zones where their extraction, transit, and sale underpin the financing of these conflicts—body parts for operations, rare animals, and rare or outlawed animal products. There are also extensive transnational trades in the smuggling of licit products such as cigarettes and even mundane household goods such as cleaning fluids where it is possible for criminal entrepreneurs to exploit price and tax differentials that open up across proximate national borders (Hornsby and Hobbs, 2007). Beyond this though, organized crime is also defined as including the provision and control of illegal or illicit services such as prostitution and gambling (in certain jurisdictions such as the United States and India, where it is prohibited or access to it restricted), cybercrime, robbery, maritime piracy, kidnapping, extortion, corruption, numerous environmental crimes, and money laundering.

Levi, though, has long argued that organized crime is a conceptually inadequate term because it covers such a complex and diverse empirical terrain, something evident from a glance at the list above. It has been argued that there is far greater diversity *within* the activities that fall under the banner of organized crime than there are *between* organized crime and other forms of criminal activity (Levi, 2002: 887, 2014: 6). Levi has argued that a more appropriate definition of the activities that are commonly regarded as constituting organized crime is the "organization of serious crimes for gain" (Levi, 2002, 2014). He also cites Naylor's (2002, 2004) useful distinction between predatory crimes, such as fraud or cybercrime; market-based crimes, which include the production and distribution of illegal goods and services; and commercial crimes, which are committed by otherwise legitimate entrepreneurs or corporations (2002: 888–889). However, even here these categories should be considered somewhat fluid and not absolute and discrete. For example, commodity counterfeiting could be considered predatory in that it involves the theft of intellectual property. It is also certainly a market-based crime as it involves the sale of counterfeited commodities. In addition, though, it may also be a commercial crime where corporations have been shown to be complicit in the acts of counterfeiting as has been suggested in the case of haute couture and other fashion counterfeiting (Saviano, 2008). There is evidence here, for example, that high-quality counterfeiting has been viewed as a form of brand promotion by some fashion houses whose complicity has been evident in a deliberately low level of policing of these copies (Harbi, and Grolleau, 2008; Wall and Large, 2010: 1104). We can also recognize differences in definitions of organized crime that stem from varying epistemic and official perspectives, all of which emphasize different aspects of the activities to which they refer. For example, some conceptualize organized crime as a set of illicit economic activities (Carter, 1997; Dick, 2009; Ruggerio, 2009; Smith, 1980), whereas others see it more in terms of social organization (Albini, 1971; Ianni and Ianni, 1972; Lombardo, 1997), while still others emphasize its complexity and the interrelations between the economic, political, social, and cultural worlds (Santino, 2016).

It is possible to detect within a number of influential definitions of organized crime reference to aspects of the organizational character of criminal groups (see Holmes, 2016: 6–9). For example, Abadinsky's (2013) definition, the product of an evolution across many editions of his text *Organized Crime*, which was first published in 1981, cites the presence of hierarchical structures, limited or exclusive membership, a unique subculture, a division of labor (within earlier editions), continuation through time, monopolies within markets, and sets of rules and regulations by which they are governed (Holmes, 2016: 7). Other well-circulated definitions, such as that by Finckenauer (2005: 5–9; in Holmes, 2016: 7), cite discernible, if increasingly less, hierarchical structures; continuity; restricted membership; and social bonds between group members. There is much within these endeavors that bears considerable empirical scrutiny. Stephenson's (2015) interview-based analysis of Russian criminal groups makes much, for example, of the social reproductions of these organizations and the importance of the gang as a social as well as an instrumentally criminal unit. However, such definitions as the ones quoted above inevitably underplay both the temporal dynamism and the spatial complexity of organized crime, something the authors of these definitions allude to with varying degrees of explicitness. It is easy to find limitations within the definitions above by referring, for example, to the wealth of empirical analysis that disputes the argument that criminal organizations display hierarchical organizational structures (Hobbs, 2001; Kenney, 2007; Kleemans and van de Bunt, 1999; Wright, 2006). Indeed, this appears to be becoming less and less the case, as will be discussed later in this book. Furthermore, while the characteristics outlined in these and other definitions might apply to *some* groups at *some* times in *some* places, it is not safe to assume that they are universal. There is enormous variation in the nature of criminal organization globally (Levi, 2002; Lombardo, 1997) and the analysis of this variation is something that has been underplayed traditionally. Criminal groups are, as this book demonstrates, somewhat more complex, dynamic, and parochial than is often supposed within cultural and political discourse (Hobbs, 1998, 2013; Hornsby and Hobbs, 2007: 558).

Cultural, political, policing, and, at times, academic discourses of organized crime have frequently evoked the specter of the existence of large, often transnational criminal mafias (Stephenson, 2015: 116). These were, perhaps, at their most extreme in cinematic cold war fantasies such as the James Bond movies and *The Man from U.N.C.L.E.* whose heroes found themselves routinely pitted against evil, inevitably Eastern European, criminal geniuses who headed shadowy, highly organized, global enterprises bent either on world domination or destruction (Dittmer, 2010; Dodds, 2005). However, such myths are not restricted to the cinema screen. Various U.S. government commissions into organized crime across the mid-20th century spoke of the presence in America of centralized Mafia conspiracies and nationwide crime syndicates, which they typically identified as being of foreign origin. Central to this imagination was the book *Theft of the Nation* (1969) produced by eminent criminologist Donald Cressey, which he based on his report for President Lyndon Johnson's Commission on Law Enforcement and the Administration of Justice (Woodiwiss and Hobbs, 2009: 111–112; see also Woodiwiss, 1988). Although highly influential in terms of the perception of criminal organization, Cressey's work has not been well received by more recent commentators who have argued it had little basis in empirical reality (Hobbs and Antonopoulos, 2013: 35; Kleemans and van de Bunt, 1999; Woodiwiss and Hobbs, 2009). Kleemans and van de Bunt (1999: 29), for example, drawing on systematic analysis of data from 40 closed investigations of criminal organizations in the Netherlands from 1996 to 1997 that were convicted of the trafficking of various drugs, of women for sexual exploitation, and of illegal immigrants; fraud; money laundering; and a host of other illegal activities, concluded that "the criminal associations that we have analysed, turn out to be less hierarchical, less stable, and far more 'fluid' than Cressey's well-known bureaucratic model suggest." Some more recent interventions within the criminological literature have again raised the specter of extensive, transnational criminal collaborations, integration, and even mergers between formerly discrete Sicilian, American, Russian, Colombian, Japanese, and Chinese organized crime groups (Jamieson, 1995; Robinson, 2002; Shelley, 2006; Sterling, 1994). Such claims have been treated with skepticism by many critical commentators (Wright, 2006: 159–160) and evidence to counter their thesis is readily available (Varese, 2011).

Thus, it is has been an enduring aspect of the criminological literature to attribute greater size and structural coherence to criminal organizations than exists empirically. Cressey's bureaucratic model of organized crime has generated a persuasive language through which to talk of the structure and organization of criminal enterprises as evidenced by the tendency of the definitions of organized crime to see structural coherence within criminal groups. Such myths of criminal organization have arisen for three reasons. First, they both fed easily into American cold war paranoias and built upon earlier fascinations that associated immigration and transgression. These became ready staples of Hollywood fantasies at the time (Hobbs and Antonopoulos, 2013: 35; Woodiwiss and Hobbs, 2009). America, in the mid-20th century, appeared ready to believe that as well as the external threat of the Soviet Union, they faced the equally serious internal threat of a criminal conspiracy emerging from within its immigrant communities. Such myths sat easily with America's views of the world at the time. Second, the bureaucratic model of organized crime posited that criminal organizations mirrored the structures of modern institutions (Hallsworth, 2003: 111, cited in Stephenson, 2015: 116). Hallsworth (2013: 112) argues that "the organisation of informal organisations cannot be grasped through imposing upon them the bureaucratic properties of formal organisations such as corporations or armies." However, this tendency provided an understandable, digestible narrative for its audiences in political, policing, public, and some academic circles and invited certain responses and reassurances, such as the view that the collapse of these organizations might plausibly follow from the taking down of their bosses. Finally, until relatively recently, there were few substantive ethnographic encounters with criminal groups available with which to counter these bureaucratic claims and insert more credible alternatives in their place (Stephenson, 2015: 116).

Despite the complexities outlined above, official and popular discourses of organized crime undoubtedly carry great rhetorical power. It has been argued, for example, that they perpetuate a series of "false discontinuities" between organized and other forms of crime and organized crime and opaque and questionable corporate and state practices (Reuter and Rubinstein, 1978; Smith, 1980, in Levi, 2002: 881). Such false discontinuities promulgate the particularly American mapping of organized criminality, discussed above, as a mobile threat, external to "normal" society (Hobbs, 1998, 2001; Madsen, 2009; Varese, 2011; Watt and Zepeda, 2012; Woodiwiss and Hobbs, 2009). Such discourses suppress suggestions of contact and interdependence between licit societies and the worlds of organized crime. Organized crime, though, depends on multiple connections to and overlaps with aspects of licit societies and economies and typically the identities of actors and activities operating within organized criminal markets are only rarely exclusively criminal in nature (Hobbs, 2013; Nordstrom, 2007). Thus, delimiting boundaries between the criminal and noncriminal becomes a problematic endeavor both empirically and conceptually (Abraham and van Schendel, 2005; Bhattacharyya, 2005; Brown and Cloke, 2007; Wilson, 2009). Furthermore, the false discontinuities identified in these critical literatures are maintained through official discourses of organized crime that isolate "essential characteristics" derived wholly from the activities and characteristics of illicit actors while overlooking empirically similar, and in many cases more serious, activities carried out by formally licit actors. These latter activities include, for example, white-collar crime, state crime, and various opaque aspects of corporate, institutional, and state practices (Aas, 2007; Cribb, 2009; Nordstrom, 2007; Reuter and Rubinstein, 1978; Smith, 1980; Tilly, 1985). As Ruggiero argues, "Transnational organised crime is not [to be] exclusively identified with the illegal activities of notorious large criminal syndicates. Rather, . . . transnational crime may well transcend conventional activities and mingle with entrepreneurial and, at times, governmental deviance" (2009: 119).

This book, then, aims to address the failure of geographers, particularly economic and, to some extent, political geographers, to recognize in any sustained way that organized crime constitutes a phenomenon worthy of geographical study, and equally the failure of others to see organized crime in geographical terms. In doing so, though, it continues to acknowledge the problems of defining and delimiting organized crime as a discrete field of practice and hence of study. It has sympathy, for example, with scholars who argue that the distinctions between those empirical practices that are conventionally defined as constituting organized criminal markets, and many opaque and outright hidden practices of business and government, are ones not of type, but of degree. This book should be seen as situated within a set of emergent geographical projects that have aimed to more clearly reveal opaque and hidden economic practices of all kinds, including those of corporate and state, as well as criminal, actors (Brown and Cloke, 2007; Gregson and Crang, 2016).

However, to provide a degree of analytical focus to this discussion, this book examines, primarily, activities that typically, within official discourses, constitute typologies of organized crime (Albanese, 2005: 10). The justification for this approach is the argument that these activities represent empirically significant sets of markets that have been largely ignored within geographical analysis and discussion of the contemporary global economy until some recent efforts have acknowledged and sought to address this lacuna (Hall, 2010a, 2010b, 2013; Hudson, 2014). Despite the inadequacies associated with the term, then, this book does use "organized crime" throughout or associated phrases such as "organized criminality" or "organized criminal entrepreneurs/groups/markets, or economies." This is not to deny the importance of recognizing the weaknesses of the term and the concepts it refers to, discussed above. This choice of terminology is in part a reflection of the extent to which it has become embedded in cultural, political, legal, and economic discourses. As Levi argues, "Whatever academics and many thoughtful practitioners may think, the term 'organized crime' may have become so culturally and legally embedded that we cannot eliminate it, despite its manifest serious defects to anyone who wishes to think analytically" (2002: 887).

ECONOMIC GEOGRAPHY PERSPECTIVES

Organized criminal markets are important parts of the contemporary economic scene. Chapter 2 provides an overview of contemporary organized crime across the world and notes the extent to which these markets generate significant revenues to nations and regions. They are an aspect of the global economic scene that deserve greater attention from academic researchers, particularly economic geographers, than they have received to date. As many authors have recognized, there are a wealth of contemporary cultural and political discourses that have offered distorted, glamorized, and hysterical renditions of organized crime (Chibnall, 2009; Hobbs, 2013; Hobbs and Antonopoulos, 2013; Larke, 2003; Rafter, 2006; Shadoian, 2003). One of the purposes of academic accounts is to counter the readings of organized crime that have tended to prevail, all be they unlikely to gain as much purchase within the social and political imaginations as these other, more easily consumed, cultural discourses.

There is no single, unitary economic geography perspective that this book and the endeavor that underpins it are able to easily slip into. Economic geography is a subdiscipline of human geography, which evolved from within British colonial and German economic traditions of the 19th century. Over the course of its history, it has embraced a number of widely divergent philosophical, theoretical, methodological, and empirical concerns and perspectives and now offers a very diverse terrain of enquiry (for a concise overview of the history and evolution of economic geography, see Aoyama, Murphy, and Hanson, 2011). Despite this diversity, though, there are sets of concerns that have remained central to economic geography perspectives despite the specificities of individual approaches. These include a concern with the geographically uneven unfolding of economic processes and their impacts across space and a recognition of the agency of uneven development in shaping the economic trajectories of territories of various scales from the local to the global (Hudson, 2014: 777). Thus, Aoyama and colleagues (2011: 1) argue:

> Contemporary economic geographers study geographically specific factors that shape economic processes and identify key agents (such as firms, labour and the state) and drivers (such as innovation, institutions, entrepreneurship and accessibility) that prompt uneven territorial development and change (such as industrial clusters, regional disparities and core–periphery).

These concerns underpin much of the analysis within this book. Specifically, it emerges from within a broadly political-economy perspective. A political-economy approach "recognises the complexity of the economy as constituted by labour processes, processes of material transformation, and processes of value creation and flow in specific time/space contexts" (Hudson, 2005: 13). This approach is attentive to the roles of a range of key agents in these processes including organized criminal groups, civil society, policing, security and intelligence agencies, corporations, states, and international institutions in the production, consumption, and mobilities of illicit and illegal goods and services (see, primarily, Chapters 4, 5, and 6). This, in itself, offers only limited innovation within the multidisciplinary literatures of organized crime, a significant proportion of which is sympathetic to, if not explicitly articulated through, broadly political-economy perspectives. Rather, its innovation comes from its explicitly geographical rendering of this political-economy approach which has, to date, not been applied in any systematic or extensive way to illicit and illegal economic processes and their outcomes. It incorporates into its analysis the relationships between geographically specific factors and the uneven contours of organized criminal markets, which operate across a variety of scales (see Chapter 5). Acknowledgment of the geographies of organized crime also highlights the potential of developing comparative analysis across different regional contexts, something that has been relatively underdeveloped within the research literature. In recognizing the social and discursive production of licit- and illicitness, it also aims to offer a culturally sensitive political economy advocated by Hudson (2005: 15).

Furthermore, the approach of this book is multiscalar in nature and attentive to the interplay of processes operating at different scales. Here it stands in contrast to the majority of previous literatures on organized crime, little of which has exceeded a single scale of analysis. It picks up also on a major preoccupation of economic geography, which is a recognition of the significance of networks of various kinds, connections between spatially distant locations, and the influence they have on the development of regions, and also methodological and conceptual approaches that take the network, rather than spatially bounded units of territory, as their site of analysis (Dicken, Kelly, Olds, and Yeung, 2001). Here again this approach differs from the majority of previous literatures on organized crime, much of which is based on regional case studies (see Chapter 5). Finally, the approach of this book is also sensitive to poststructural insights, which have gained increasing traction within some literatures of economic geography. Here we are concerned particularly with the discursive framing of space and of movement and the ways in which this conditions official responses to the production and consumption of illicit and illegal goods and services (see Chapter 6). Such responses are themselves key factors in understanding the emergence, endurance, and change within organized criminal economies. Ultimately, it is hoped, that the perspective outlined in this book can highlight pathways to more nuanced, plural, and effective responses to the problems of organized crime (see Chapter 7). While undoubtedly expanding the empirical range of economic geography, this book highlights a number of issues that have been underappreciated within the literatures of organized crime thus far.

The economies of organized crime embody many issues that have received wider attention from economic geographers and which they have argued are central to understanding the natures of contemporary global economic formations. These include transnational networks (Coe, Dicken, and Hess, 2008; Dicken et al., 2001), diverse economies, and alternative economic spaces (Fickley, 2011; Gibson-Graham, 2008; Leyshon et al., 2003) and opacity and risk in economic systems (Cobham, Janský, and Meinzer, 2015; Wójcik, 2013). In this sense, these economies and their spaces are amenable to examination from the perspectives of economic geography and likewise economic geography is conceptually and methodologically well equipped to contribute to this endeavor (Hudson, 2014: 777).

In the first instance economic geography could profitably recognize that its discourses are incomplete given the silences identified above. There are economic geographies of illegal markets, then, such as narcotics, weapons trafficking, people trafficking, smuggling, counterfeiting, money laundering, cybercrime, corruption, maritime piracy, and environmental crime, among many others, that remain almost entirely unwritten as yet. Furthermore, though, criminality and corruption are not restricted to these formally illegal realms. Rather, they are present within many economically mainstream processes, networks, and markets such as the international trades in secondhand cars and fashion explored through Brooks's (2012, 2015) innovative research. While economic geographers have advanced understanding of many of these economies they, again, have said little, with the exception of work in the vein of Brooks's cited above, about the presence, roles, and influences of criminal and corrupt actors and activities within them.

There are undoubtedly many overlaps between licit and illicit, legal and illegal economies and their spaces. Thus, economic geography, conceptually and methodically, is well placed to engage with the geographies of illicit and illegal economies. However, there are also many differences between these realms that suggest more nuanced analysis. Naylor, for example, is critical of the lazy representation of criminal organizations as businesses that just happen to be operating in markets that are illegal in nature:

> The facile analogy between legal and illegal firms is at best an oversimplification, at worst simply wrong. In illegal markets that are highly segmented, decisions are personalized, information flows constricted, capital supplies short term and unreliable, objective price data lacking, and the time horizons (indeed, the very existence) of enterprises coterminous with those of the entrepreneurs. (2004: 21)

Varese's (2011: 190–192) work on mafia migrations from Italy, although written not from the perspective of geography but rather from that of sociology/criminology, has demonstrated that such groups do not act in the same ways as legitimate global corporations, rationally seeking out opportunities across international space. Rather, their mobilities are more forced and their successes more grounded in local conditions than is perhaps the case for legitimate transnational corporations. Furthermore, Weinstein (2008) has shown how the liberalization and globalization of the Indian economy in the 1990s, a time of economic boom generally, albeit one whose benefits were distributed in markedly uneven ways, actually impacted negatively on the income streams of organized criminal groups there, leading them to embed themselves in alternative markets. Thinking of these criminal groups as paracorporations, then, might be misleading, which questions the simple transferability of economic models developed in the study of licit economies to the illicit and illegal. This point is picked up particularly in Chapter 4, which looks at the organization of criminal groups. However, there is no reason to think that economic geography cannot contribute to the production of spatialized readings of these realms. The extant literatures of organized crime have long been characterized by a multidisciplinarity to which economic geography has contributed little to date. There is little reason, though, to imagine that this omission should continue to remain unaddressed.

THE POLITICAL AND SOCIAL PRODUCTION OF ORGANIZED CRIME

As the discussions above and throughout this book illustrate, crime, and, if we think specifically of the concerns of this book, organized crime, are produced by state actions that criminalize certain acts and not others, which vary through time and from country to country. Leaving aside, for the moment, the question of whose interests are served by these precise definitions of legality and illegality (though see Hudson, 2014: 777), there are many examples of these multiple contingencies that are pertinent to the discussions in this book. Discussing Andreas's (2002) work, Madsen (2009: 2), for example, argues "that a century ago, most of the crimes now relating to smuggling were at most revenue violations (drugs, endangered species, etc.), which have been criminalized by the expansion of so-called prohibitionist laws." Such laws have, of course, generated lucrative markets in prohibited commodities that have been fundamental to the evolution of organized crime internationally since the mid- to late 20th century (LSE Expert Group on the Economics of Drug Policy, 2014; McCoy, 2004; Watt and Zepeda, 2012).

Recently we have witnessed some movement away from this universal prohibition of drugs with the decriminalization of the possession of certain quantities of certain types of drug, in some cases only for certain people, such as the seriously ill in the case of medical marijuana/ cannabis, in places such as Germany, the Netherlands, Portugal, Colombia, Costa Rica, Mexico, Uruguay, and the U.S. states of California, Colorado, and Washington, among others. Thinking about crime, then, of any kind, is an inherently spatial process, even if it is commonly not acknowledged as such, as it involves, implicitly at least, consideration of the context (legal, but also economic, political, and social/cultural) that produced that criminal act through demarcation in law.

These contingencies are further complicated if we roll legal statues together with the realm of social and cultural custom and practice. At times, in many places, legal statutes that formally define acts as criminal clash with social custom and practice that deem these same acts as socially licit. The result here is that often social and cultural custom and practice trump legal statute. Examples include the cultivation and consumption of coca leaves, from which cocaine is manufactured, in parts of Latin America such as northern Argentina ("Here even legislators chew them"; Cusicanqui, 2005: 128) and the significant financial benefits that China derives from the smuggling of a host of counterfeit commodities (Hudson, 2014: 786; Phillips, 2005) and people (Glenny, 2008: 366–367; Scott-Clark and Levy, 2008), which has led to these activities receiving the tacit approval of some provincial governments in the absence of, or only limited, policing through a form of "local protectionism" (Chow, 2003: 473;

Glenny, 2008: 366–367; Lintner, 2005). Thus, while activities such as these might be formally illegal, under certain conditions they may emerge as socially licit practices (Abraham and van Schendel, 2005; Nordstrom, 2010: 173) and ones that officials are prepared to turn a blind eye to. This discussion hints at distinctions between law, policing, and practice, whereby some activities, while deemed formally illegal, are routinely overlooked by governments and law enforcement agencies and accepted within the social and cultural practices of wider populations. As was the case with the formal legal contingencies discussed above, complex spatial contingencies emerge again across the terrain of social custom and practice.

We can attempt to think of the relationships between the legal/illegal and licit/illicit schematically by considering acts that are legal and licit (many that characterize the practices and spaces of the formal economy); acts that are legal but illicit (such as some forms of aggressive tax avoidance); acts that are illegal but licit (such as the two cases discussed in the previous paragraph); and acts that are illegal and illicit (such as physically, symbolically, and economically violent activities like extortion, contract killing, or armed robbery) (based on Chiodelli, Hall, Hudson, and Moroni, 2017; see also Abraham and van Schendel, 2005: 20). Over such ideal-type classifications, though, we must again layer considerable spatial, temporal, and social complexity, contingency, and qualification. Although useful, it is impossible to construct a categorical framework such as this that can be applied universally. Activities may shift categories as we move around the world. For example, while people trafficking is universally prohibited in law and generally considered an abhorrent, illegal, and illicit activity, this is not necessarily the case everywhere. The example in the previous paragraph shows that this social judgment does not necessarily apply among some members of the populations and government officials, at least, of those Chinese provinces heavily implicated in this illegal trade in people. Here pockets of opportunity may open up for organized criminal actors in such spaces. Thinking through the production of illegality and illicitness, then, is an inherently spatial exercise as, to be meaningful, judgments of these types must refer to the economic, political, legal, and social/cultural contexts within which activities take place. Similarly, nuanced understandings of the illegal and illicit economies that such social and legal contingencies help to sustain in some spaces, clearly including those of organized criminal activities, should be equally spatially informed.

PRIOR GEOGRAPHICAL LITERATURES OF ORGANIZED CRIME

While illicit and illegal activities associated with organized crime have not been totally overlooked by geographers, their engagement with these realms has been both limited and somewhat disparate to date. Taylor and colleagues' (2013) review is revelatory in that it excavates a somewhat overlooked and more extensive tradition of geographers' engagements with illegal economies, specifically those associated with narcotics, than is generally acknowledged within disciplinary discourses. This work speaks of a tradition of geographical analysis that can be traced back to the 1980s, which, across a range of subdisciplines, have outlined the rich geographies of narcotics. Their discussion acknowledges geographical contributions from social, cultural, health, rural, political, environmental, and ecological geography, although there is relatively little from, primarily, economic geography perspectives. The works discussed have revealed aspects of the geographies of narcotics production, agricultural, ecological and cultural impacts, consumption, addiction and treatment, policing and prohibition, and their roles in conflict and development, state formation, and urban restructuring. Despite the wealth of geographical work identified here, these contributions, pioneering in many instances, have made relatively little impact on broader disciplinary debates to date.

One of the most fully realized geographical reflections on narcotics, and one that Taylor and colleagues (2013) discuss in their review, is the collection *Dangerous Harvest: Drug Plants and the Transformation of Indigenous Landscapes* (2004) edited by Steinberg, Hobbs, and Mathewson. This is a collection of essays, global in their collective scope, the majority of which are authored by academic geographers or academics and practitioners from very cognate, predominantly life science and environmentally oriented disciplines such as anthropology, biology, environmental design, history, and medicine. The collection is innovative and noteworthy in three senses. First, all of the essays are written through an explicitly geographical lens, rare for work on this subject. The authors' perspectives are broadly cultural/moral geography in orientation. Second, the authors all explore narcotics, not from the positions that have tended to dominate debate such as those of international policy, law enforcement, or the criminal entrepreneur, but rather from the perspectives of indigenous people for whom drug plants have long played vital historical and cultural roles. These are voices largely absent from prevailing drug literatures. It is the changing relationships between indigenous people and these plants, as the latter have become increasingly central to transnational political economies of drug trades and their associated policing and eradication efforts, that the authors are most insightful on. Third, the global scope of the essays in the collection is revelatory of both the extent of the issues the book discusses and of their geographical and historical contingencies. *Dangerous Harvest,* then, demonstrates the originality and richness that geographical perspectives can bring to discussions of the illicit and of the commodities of illicit trade and their effects. However, in relation to the key concerns of this book, organized criminal actors and the economic dimensions of the drug trades that are discussed there are less central concerns.

It is possible to discern tentative signs of a wider emergent interest within geography in aspects of illicit and illegal economies, though. For example, it is not unknown for geographers to acknowledge the presence or significance of criminal or corrupt actors within various political, social, and economic contexts (Hudson, 2005: 6; 2016; McNeill, 2004: 127–128; Perrons, 2004: 325). However, this has rarely resulted in any substantive associated engagement. Among the few exceptions to this are Rengert's (1996), Allen's (2005), and Chouvy's pioneering geographical/cartographical analyses of the production, distribution, and consumption of narcotics. Chouvy, for example, has produced a substantial body of work on the Moroccan hashish industry (2005a, 2005b) and opium production in the Golden Triangle and Golden Crescent (2009) and the wider trafficking economies of Southeast Asia (2013). Here geographical models and spatially sensitive perspectives, more usually applied to licit economies, have been deployed to read the spatialities of narcotics economies. However, they have achieved only limited purchase within geography's literatures. Rengert's work, for example, has been more widely cited within the criminological, crime prevention, and environmental psychology literatures (Curtis and Wendel, 2000; Ratcliffe, 2004; Taylor, 2003) than within geographical circles. In the years immediately following their release Rengert's and Allen's works were largely overlooked by geographers. They have been cited more frequently recently as researchers, writing within geographical discourses, have begun to more readily look beyond the licit (Corva, 2008; Hastings, 2015; Pereira, 2010). In addition to Rengert's and Allen's empirical engagements, Deverteuil and Wilton's (2009) review also speaks to the geographies of drug production, consumption, regulation, and treatment, although from a social rather than from an economic geography perspective.

Corruption, in development, post-Soviet, and Northern contexts, is also an issue that has drawn attention from geographers in various ways (Brooks, 2012, 2015; Brown and Cloke, 2004, 2007; Swain, Mykhnenko, and French, 2010). Brown and Cloke's work has called for geographers to recognize the significance of illicit and illegal transnational financial movements as central elements of contemporary neoliberal globalization that have been insufficiently acknowledged within geographical accounts. Brooks's analysis of transnational commodity chains in the used-car, fast fashion, and secondhand clothing industries highlights the ways in which neoliberal discourses have obfuscated criminal presence, corrupt practices, and opacities within markets that are often painted as development successes. Brooks's work is returned to again in more detail in Chapter 6 of this book.

Finally, there are a series of emergent contributions that have either sought to map the geographical literatures on illicit and illegal economies or to open up economic and political geography to discussions of the illicit and illegal (Chiodelli, Hall, and Hudson, 2017; Hall, 2010b, 2012, 2013; Hudson, 2014, 2016). There are, then, some accounts within geography that acknowledge the illicit and illegal as legitimate objects of scrutiny and within this body of work some that have engaged with organized criminality directly as a series of industries comparable in their extent to many licit industries that have long drawn the attention of geographers. However, it is clear that economic geographers to date have shown relatively little appetite to pursue these agendas empirically. Geography, though, seems more open now to dialogues around these issues than has been the case at many points in the past.

THE BOOK

It is worth finishing by outlining five of the key questions that this book addresses.

- What are the spatial configurations of organized criminal activity globally?
- How can the spatial patterns of organized crime that are observable at a variety of scales be explained?

- In what ways are these geographies of organized crime active within ongoing processes of economic change?
- How can the geographies of organized crime be understood and interpreted empirically and conceptually?
- In what ways can geographical interpretations of organized crime be applied in responses to organized crime?

In some cases the questions above infuse multiple chapters. The first four of these questions are present in many of the discussions in Chapters 4, 5, and 6, which consider, in turn, the organization of criminal economies, the spatialities of organized crime, and the patterns of criminal mobilities and their representation within official discourses. Chapter 4 looks at the organizational structures of criminal groups currently operative within criminal economies. It is concerned initially with exposing the flaws of some cultural myths of criminal organization. While attempting to convey something of the empirics of criminal organization today, the chapter attempts to do so through an explicitly geographical lens that raises questions such as the extent to which place offers an enduring basis for criminal organization and the ways in which forms of regulation deployed within the criminal economy might restrict the extent of specific criminal hegemonies. The central concerns of Chapter 5 are the ways in which conceptual understandings of the empirical, observable distributions of organized crime might be developed. Here the chapter considers questions of scale in interpreting patterns of organized crime along with more networked ontologies. In moving through this discussion, it considers a number of alternative theoretical frameworks that emerge out of these perspectives. Chapter 6 is concerned both with actual patterns of criminal mobility, particularly those associated with the transit of illicit and illegal commodities across transnational space, and with their representation within official discourse. It attempts to blend something of the empirics of contemporary criminal mobilities, conceptual understandings of them, and a critical unpacking of their renditions within official discourses and the effects of this. It is particularly concerned with conveying a sense of the groundedness of these movements, which tend not to be captured in official cartographies. The final of the five questions outlined above, although referred to at a number of points within the book, is considered in detail in Chapter 7, which looks at extant responses to organized crime, their limitations and flaws, and the potentials for more nuanced, geographically sensitive, policies. Before that, though, the book moves to two chapters that provide some contextual discussions. Chapter 2 offers an overview of recent patterns and trends in organized crime globally, while Chapter 3 considers approaches that have been employed to measure and research these organized criminal economies.

CHAPTER 2

CONTEMPORARY ORGANIZED CRIME

INTRODUCTION

In aggregate terms the markets associated with organized criminal activities constitute a significant proportion of contemporary global economic activity. Various estimates place the value of these markets at between US$870 billion and US$1.3 trillion per annum or between 7 and 20% of global GDP (gross domestic product) (Glenny, 2008; Lemahieu, Sampaio, & Comolli, 2015: 2; United Nations Office for Drugs and Crime, 2015c: 1; World Economic Forum, 2011: 23). A significant proportion of the variation in published estimates is accounted for by methodology and whether illicit financial flows associated with money laundering are included or excluded from these figures. Regardless of any notes of caution associated with estimates of these economies, which are returned to in some detail in Chapter 3, any reading of global economic activity that does not acknowledge the illicit and illegal economies of organized crime is inherently partial in its coverage. In some transitional and developing countries criminal markets have accounted for up to 40 to 50% of national income (van Dijk, 2007; Dunn, 1997; Glenny, 2008). There are even the occasional outliers beyond this, such as Angola in Central Africa, where 65% of its economy has been described as "extralegal" (Nordstrom, 2007: 5). It is clear, then, that "in some countries, illicit trade is the major source of income" (World Economic Forum, 2011: 23; see also Nordstrom, 2007, 2010: 173). However, extensive criminal economies are not the exclusive preserve of fragile or emergent economies. The European Union and the United States, for example, both sustain significant criminal markets across a range of sectors including, but not limited to, illegal drugs, human trafficking, firearms trafficking, cybercrime, counterfeiting, illicit trade in tobacco products, fraud, and corruption (Europol, 2013; Ferragut, 2012; Glenny, 2011; Transcrime, 2015; United Nations Office on Drugs and Crime, 2014a) that both generate economic revenues for criminal entrepreneurs and pose complex issues for the policing, policy, security, and civil communities that they have long failed to address in adequate, let alone effective, ways (see Chapter 7).

Criminal entrepreneurs infiltrate and exploit the architecture of mainstream economic globalization including onshore and offshore financial institutions, information technology, and global logistics, as well as multiple licit economic sectors (Deneault, 2007; Ferragut, 2012; Glenny, 2008, 2011; Kilma, 2011; Nelen, 2008; Nordstrom, 2007; Urry, 2014; Varese, 2011). It is wrong, then, to paint organized criminal economies, as many cultural and political discourses do, as exogenous threats or as empirically or ontologically discrete and beyond the economic and political mainstream. Multiple literatures have persuasively demonstrated the ways in which these overlap and intersect with these mainstreams in diverse ways in both core OECD (Organisation for Economic Co-operation and Development) economies and in those of the global South (Bhattacharyya, 2005; Brown and Cloke, 2007; Castells, 2000; Hall, 2013; Hudson, 2014; Nordstrom, 2007, 2010). Furthermore, the governance and development roles played by organized criminal groups have been recognized in a variety of regional, economic, and political contexts (Goodhand, 2009; Madsen, 2009; Ruggiero, 2009; Weinstein, 2008; Wilson, 2009; Wright, 2006: 144), although the contributions of markets associated with organized crime have been contested by many authors (see Jesperson, 2017; van Dijk, 2007), a point returned to later in this chapter.

This chapter provides a brief overview of contemporary organized crime. It aims to be global in extent and to blend discussion of general patterns and trends that have evolved, broadly since the late 1980s, with some discussions of specific criminal economies and organized crime within various key regions. The chapter does not consider in any depth histories of organized crime prior to this period, which is somewhat beyond the scope of this book and is well covered, in different ways, by other authors (see Galeotti, 2008, and the other contributors to this special edition of the journal *Global Crime*; see also Hobbs, 2013; Knepper, 2009, 2011; Wright, 2006). The aims of the chapter are to provide both a survey of the international landscape of contemporary organized crime and to foreshadow a number of issues that are discussed in more detail later in the book.

THE ECONOMIES OF ILLEGAL DRUGS

Of contemporary organized criminal markets, undoubtedly the largest in aggregate global terms is the production, distribution, and sale of illegal drugs. Notwithstanding the concerns about the nature and accuracy of data and measures of illicit and illegal markets that are discussed in Chapter 3, it has been estimated that narcotics markets generate roughly US$320 billion annually (Table 2.1) and represent, according to different estimates, between 35 and 85% of the global income of organized criminal groups (Galeotti, 2005a: 1; Glenny, 2008: 262; Midgley, Briscoe, and Bertoli, 2014; 5; United Nations Office on Drugs and Crime, 2015c: 1). The narcotics trade has been estimated to be equivalent to 1.5% of all money moving through the world financial system (United Nations Office on Drugs and Crime, 2011, cited in Taylor et al., 2013: 415) and 7–8% of world trade. The latter figure suggests that it represents an industry of comparable size to the global textile industry (Moynagh and Worsley, 2008: 176; see also Bhattacharyya, 2005; Wright, 2006; Madsen, 2009).

Looking inside these aggregate figures, cannabis is the most widely consumed drug globally, with an estimated world market worth of US$140 billion (World Economic Forum, 2011: 23). The

TABLE 2.1. Income Generated by Transnational Organized Crime in Billions per Year (US$)

Illicit drugs	320
Illegal arms	250
Counterfeit	250
Human trafficking	32
Global business worth	870

Note. Based on United Nations Office on Drugs and Crime (2015b).

cocaine and opiates markets have been estimated at US$85 billion and US$68 billion, respectively, in 2009 (United Nations Office on Drugs and Crime, n.d.), while the World Economic Forum reported estimates of their worth at US$80 billion (cocaine) and US$60 billion (opium and heroin) (World Economic Forum, 2011: 23). Figures from the European Union suggest sales of illegal drugs are worth €100 billion per annum (figures from Europol, cited in World Economic Forum, 2011: 3) and 2010 estimates for the United States suggest an internal drugs market worth in total US$108.9 billion, which is made up of markets for cannabis/marijuana worth US$40.6 billion, cocaine worth US$28.3 billion, heroin worth US$27 billion, and methamphetamine (crystal meth) worth US$13 billion (RAND Corporation, 2014). Despite the size of these markets, it is generally agreed that the scale of cocaine and opium/heroin production has declined since the 1990s; however, globally, the production (and consumption) of cannabis/marijuana appears to have grown during the early years of the 21st century (Wright, 2006: 79–80).

The economies associated with individual drugs display radically different geographies. Opium poppy/heroin, for example, shows a highly concentrated pattern of production, with Afghanistan alone accounting for over two-thirds of the world area under illicit opium poppy cultivation (United Nations Office on Drugs and Crime, 2008, 2015c: xiv, 2016: xii). Outside of Afghanistan, cultivation of opium poppy and production of heroin also occurs in Myanmar (formerly Burma), which is the world's second largest producer of opium (Varese, 2011: 172–175), the Lao People's Democratic Republic, some other countries of Southeast Asia, and to a limited extent in Mexico and Colombia. The Myanmar–Lao–Thailand "Golden Triangle," which before the early 2000s dominated world opium cultivation, has shown a resurgence in cultivation in recent years, particularly in Myanmar (United Nations Office on Drugs and Crime, 2016: xiii) where there is evidence of illegal opium factories receiving informal protection by the army in the Wa region (Varese, 2011: 173). Mexican opium poppy production has shown some significant growth since 2006, where the area under production has been estimated as growing from 2,000 hectares in the year 2000 to a subsequent high of almost 20,000 hectares, making Mexico the world's third largest opium poppy grower. This growth has been fuelled by the diversion of sections of the Mexican army from poppy eradication duties in regions such as the Sierra Madre mountains to anticartel duties in northern Mexico's border and port cities, and the growth in demand for heroin in the United States following the extensive prescription of opioid painkillers by American doctors since the 1990s (Wainwright, 2016: 233–239).

Cocaine also shows concentration in the geographies of its production, with almost all of the world's production occurring in Latin America and overwhelmingly in Peru, Bolivia, and Colombia. Production patterns in these countries have demonstrated shifts in recent years in response to the extent and effectiveness of antiproduction initiatives enacted within individual countries (McCoy, 2004). This dynamism in the geographies of drug production, the so-called "balloon effect" (so called because when a balloon is squeezed down in one area it will bulge out in another) is considered in more detail in Chapter 7, which explores responses to the problems of organized crime and their limitations.

In contrast, the geographies of marijuana/cannabis production are much more diffuse and linked to specific markets (Figure 2.1) as "the drug is produced locally and consumed widely in countries around the world" (United Nations Office on Drugs and Crime, 2010b: 25; see also Wright, 2006: 80). Similarly diffuse geographies of production close to market are observable in the cases of amphetamine-type stimulants and new psychoactive substances that, albeit viewed across a short period of time, have shown some dynamism in terms of their locations of production (United Nations Office on Drugs and Crime, 2016: xiv–xv).

The reasons for such diverse narcotics geographies are multiple and complex. Undoubtedly climatic and environmental factors are important here but are far from determinate. Opium poppies, for example, can potentially be cultivated in many regions around the world, as they grow happily in temperate climates and they are a very water-efficient crop compared to, for example, wheat (Urry, 2014: 164). That they have not displayed more diffuse geographies of production highlights the significance of political and security factors in shaping the geographies of organized crime. A key factor in the growth of major opium-producing regions has been a geopolitical dimension that has overlain the macrogeographies of global drug production in the post–World War II period. It is well known and expressed across a variety of literatures that Northern superpowers have long used conflicts in distant parts of the world as proxy wars through which to pursue and extend their international geopolitical ambitions, rather than engaging in direct combat with rivals. These Northern interventions typically necessitate alliances with and

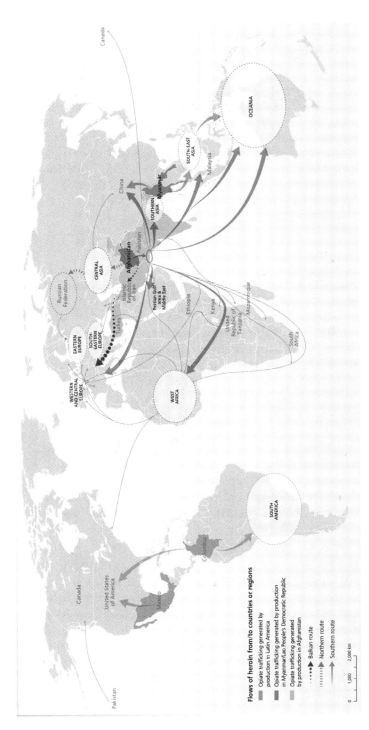

FIGURE 2.1. Main global trafficking flows of opiates. Data from United States Office on Drugs and Crime, responses to an annual report questionnaire and individual drug seizure database. The trafficking routes represented on this map should be considered broadly indicative and based on data analysis rather than definitive route outlines. Such analyses are based on data related to official drug seizures along the trafficking routes as well as official country reports and responses to annual report questionnaires. Routes may deviate to other countries that lie along the routes and there are numerous secondary flows that may not be reflected. The boundaries shown on the map do not imply official endorsement or acceptance by the United Nations. Dashed lines represent undetermined boundaries. The dotted line represents approximately the Line of Control in Jammu and Kashmir agreed upon by India and Pakistan. The final status of Jammu and Kashmir has not yet been agreed upon by the parties. The final boundary between the Sudan and South Sudan has not yet been determined. From United Nations Office on Drugs and Crime (2015: xiv). Reprinted by permission.

sponsorship of powerful local actors in return for their services and support. In many cases the source of this local power is involvement in illegal trades, of which the opium trade has been the most common and most lucrative. The involvement of the U.S. Central Intelligence Agency (CIA) has been identified as particularly significant in stimulating the development of some of the world's largest drug-producing regions. McCoy (2004: 31), for example, has argued:

> State security services have often allied with highland warlords and transnational criminals, bartering protection for their clandestine services. At the Cold War's start in the 1940s, the Iron Curtain fell along an Asian opium zone from Turkey to Thailand, drawing the United States Central Intelligence Agency (CIA) into these remote, rugged borderlands to wage covert wars that became inadvertent catalysts for the region's drug traffic. . . . A decade after the Cold War's end, the CIA's three main covert battleground's—Afghanistan, Burma and Laos—were, in that order, the world's three leading opium producers.

Watt and Zepeda (2012: 56) also identify CIA involvement in Mexico as key in facilitating the growth of the extensive cocaine-trafficking trade through the country into U.S. markets.

The basic materialities of specific drugs, when read through the lens of the economic geographies of industrial location, also affect these patterns. Heroin, the finished product, after it has been processed from opium poppies, is a high-value, low-volume drug. Cocaine, as it is processed from coca leaves, also progressively acquires these characteristics. However, while the end product might be relatively small in these cases, the raw material requirements of these illegal trades are extensive. As Boivin (n.d., n.p.) argues, "Producing small quantities of drugs [cocaine and heroin] requires very large quantities of coca or opium poppy, which means large-scale outdoor cultivation. While corruption and less intense drug law enforcement allow this kind of cultivation in some peripheral countries, the same situation is inconceivable in core countries."

Cocaine production involves, initially, the processing of bulky coca leaves in crude laboratories, or "*bossas*." These are typically found in remote rural, mountainous, forested areas located close to the sources of their raw materials. The paste that results from this process is much lower in bulk and higher in value and can be profitably, and relatively safely, transported to more sophisticated, distant laboratories in urban areas where it can be processed into the finished product (Allen, 2005; Rengert, 1996). From here cocaine is transported by now well-established routes into its main markets in Europe and the United States. These material characteristics make the finished heroin and cocaine relatively cheap to transport over long distances and allow multiple opportunities for concealment to evade detection. Thus, the location of production distant to market, in the cases of these drugs, does not impede their development into large criminal economies.

Cannabis, however, particularly in its herb form, is a relatively bulky lower-value commodity by comparison, which makes its transport across long distances more expensive and across international borders more difficult to conceal and hence more risky. Cannabis herb also has a potent and distinctive smell that is difficult to entirely mask in transit. For these reasons, and others such as more relaxed attitudes toward consumption; the development of hydroponic and lighting technologies that can be incorporated into relatively small domestic, indoor growing units; and less rigorous law enforcement efforts (Boivin, n.d.), cannabis production tends to be distributed geographically and located close to market. Custom and tradition, particularly in the cases of drugs such as cocaine in Latin American and khat in the Horn of Africa and its diaspora communities, have also historically been important to the development of relatively entrenched patterns of drug production.

The rise in popularity of synthetic drugs, though, seems to have reduced the pull of these basic economic geographies and thus production units involved in the synthesizing of narcotics are able to be more footloose in their location choices (Abdullah et al., 2014: 14). This mobility is important to the maintenance of anonymity of these drug production networks. Frequent changes of location of narcotics laboratories are common here, as is the occasional use of vehicles within which to house mobile laboratories that can near continually move to evade detection (Abdullah et al., 2014: 18), something familiar to viewers of the U.S. television drama *Breaking Bad* (2008–2013).

OTHER COMMODITY TRAFFICKING

Beyond the economies of narcotics, the next most significant criminal industries globally in economic terms are the trades in illegal arms and counterfeit goods, which have been estimated by the United Nations Office on Drugs and Crime to be roughly similar in economic worth (United Nations Office on Drugs and Crime, 2015b). Intellectual property right theft in the form of commodity counterfeiting accounts for roughly 7% of world trade, two-thirds of which originates in China (Glenny, 2008: 14; Phillips, 2005; United Nations Office for Drugs and Crime, 2010a). Other major contributors to this trade are regions such as Russia, south Asia, and Latin America, which are also significant producers of counterfeit goods (Hudson, 2014: 783).

The World Economic Forum has reported evidence of major global markets in counterfeit pharmaceutical drugs, which were estimated as being worth US$200 billion, and counterfeit electronics, estimated as worth US$100 billion (2011: 23). Both of these markets carry potentially significant public health and safety consequences. Seizure statistics from European borders suggest that the most counterfeited commodities entering Europe are clothing accessories and shoes (57% of European seizures in 2008); watches and jewelry (10%); electrical equipment (7%); medicines (6%); CDs, DVDs, and cassettes (4%); toys and games (4%); cosmetics (4%); computer equipment (1%); and cigarettes (1%) (United Nations Office on Drugs and Crime, 2010a: 178).

Although a smaller part of the global counterfeiting economy relatively, the smuggling and counterfeiting of cigarettes, to avoid duty or to exploit tax differentials between countries, has become a significant source of income for organized criminal groups (United Nations Office on Drugs and Crime, 2010a). Estimates suggest that this economy was worth US$50 billion globally in 2010 (World Economic Forum, 2011: 23) and that annual tax revenue losses from cigarette counterfeiting and smuggling to the European Union totalled €10 billion (Interpol, 2014) and that those to the British treasury were in the region of £2 billion (US$3.3 billion) per annum during the early 21st century (Hornsby and Hobbs, 2007: 553; Interpol, 2014: 31; Nordstrom, 2007: 23). These data suggest that the "share of the illicit cigarette market in the EU has continued to increase in recent years (from 8.3% in 2006 to 10.4% in 2011)" (Interpol, 2014: 43) and that roughly 25% of all cigarettes consumed in the United Kingdom, which has among the highest taxes on cigarettes within the European region and hence among the most expensive cigarettes by retail price (Tobacco Manufacturers Association, 2016), are counterfeited or smuggled (Tobacco Manufacturers Association, 2016: 31). China has been identified as a major supplying region within the contemporary trade in counterfeited brand cigarettes (Madsen, 2009: 3).

However, the criminal trade in cigarette smuggling was identified as playing a key role in the financing of the Yugoslav wars of the 1990s, and has thus long been a significant source of criminal finance in some regions (Glenny, 2008). Permissive social attitudes toward the consumption of smuggled and counterfeit cigarettes and the low risks involved compared to smuggling narcotics are undoubtedly attractive to criminal entrepreneurs engaged in these illicit trades. Furthermore, the complicity of cigarette manufacturers in the smuggling of cigarettes has been widely acknowledged (Glenny, 2008; Hornsby and Hobbs, 2007: 553; Madsen, 2009).

PEOPLE TRAFFICKING

Although significantly smaller than these trades in purely economic terms (United Nations Office on Drugs and Crime, 2015c: 1), people trafficking, predominantly from the global South for the purposes of sexual exploitation for profit and other forms of forced, informal, and illicit labor, is the most rapidly growing source of income for organized criminal groups (Foltz, Jackson, Oberg, and the European Futures Observatory, 2008; International Labour Organization, 2012; Madsen, 2009; Westmarland, 2010). It has been estimated that globally between 2.4 and 4 million people are the victims of trafficking each year. Of those affected by human trafficking, the European Union stated that "women and girls represent 56% of victims of forced economic exploitation and 98% of victims of forced commercial sexual exploitation" (European Union, 2015: n.p.).

The global revenues from people trafficking are estimated to be up to US$32 billion per annum (Financial Action Task Force, 2011: 6; Foltz et al., 2008: 8; Galeotti, 2005a: 3; Madsen, 2009: 45; United Nations Office on Drugs and Crime, n.d.; Wright, 2006: 98), although these estimates are likely to be conservative in nature. The markets of the United States and Europe form the major global destinations of human trafficking flows. Estimates for 2009 suggest that three million migrants were smuggled into the United States from Latin America, which generated US$6.6 billion, while 55,000 migrants were smuggled into Europe from Africa, generating US$150 million (United Nations Office on Drugs and Crime, n.d.). The illegal mobilities associated with people trafficking are underpinned by combinations of factors at various scales including widening global asymmetries; social and political unrest in regions of the global South; the exploitation by traffickers of cultural traditions of fostering, migration, remittance, and social norms legitimizing trafficking; tacit state approval for people trafficking where it generates significant financial movements into deprived regions; the hardening of international borders; and public demand for sex workers and cheap labor across a number of economic sectors in the global North (Aas, 2007: 36; Bhattacharyya, 2005: 159, 174; Financial Action Task Force, 2011; Glenny, 2008: 366–367; Madsen, 2009: 43; Manzo, 2005; Passas, 2001; Scott-Clark and Levy, 2008; United Nations Office on Drugs and Crime, 2014b; Westmarland, 2010).

CYBERCRIME

Undoubtedly, cybercrime is becoming an increasingly significant source of income for organized crime groups internationally. Cybercrime is a predatory form of crime and as such is not reliant on the elasticities of demand associated with the consumption of illicit and illegal goods and services. Reliable statistical measures of the extent of cybercrime are currently more elusive even than those for other forms of organized criminal activity with longer histories. The 2011 Norton Cybercrime Report estimated that the total cost of cybercrime was US$388 billion (Symantec, 2011, cited in World Economic Forum, 2012: 4). Of this cost, US$114 billion was in direct costs, which includes both criminal revenues from cybercrime and the amounts expended by citizens, corporations, and public bodies to defend against cybercrime. The remaining losses in this estimate are accounted for by time lost due to the disruptions of cybercrime. However, the World Economic Forum acknowledges that this estimate is disputed and also cites a more skeptical assessment that argues that the direct costs of cybercrime to citizens, including criminal revenues, are relatively small but that the indirect and defense costs of cybercrime are proportionally much higher (Anderson et al., 2013). Geographically, cybercriminals have displayed marked regional clusterings in Eastern Europe, particularly Latvia, other Baltic states, and Russia, all of which target external victims; Turkey; Nigeria; Brazil; and more recently China (Glenny, 2008, 2011). The factors underpinning these concentrations include international variations in the laws governing the Internet, the indifference of authorities toward local cybercriminals (Eastern Europe), computer-literate populations with few opportunities in the legitimate economy (Brazil and

Nigeria), histories of technology-based scams, and cultural hostilities toward the West that are deployed to legitimize cybercriminal activities (Nigeria).

Cybercrime clearly illustrates the social and geographical contingencies of licit- and illicitness discussed in the preceding chapter. An aspatialized reading of cybercrime might render it as a series of acts that are universally illegal and illicit and that are often seen as transcending territory by virtue of their taking place in cyberspace (though see Zook, 2003, on the naivety of ungrounded readings of the cyberworld). This is far from the case, though, as more spatially sensitive interpretations reveal. Cybercrime, then, is a series of activities that are illegal in that they are universally criminalized in law. They are also, broadly speaking, activities that are condemned socially in terms of the economic and psychological harm they can cause to victims and the threats they pose to personal, corporate, and national security, although there is some debate about the status of potentially ethical hacker groups such as Anonymous who have, for example, launched cyberattacks against the Islamic State, and Wikileaks, discussed below. The prevailing view of cybercrime, though, and certainly in its most economically predatory manifestations, is of a series of illegal and illicit acts.

This view, however, says nothing of the geographical patterning of these categories, the ways that they vary across space, and the reasons for these geographies. It is now widely accepted, for example, following the revelations of whistle-blowers such as on Wikileaks, that security agencies of national governments such as the United States, the United Kingdom, Russia, China, and North Korea, among others, have routinely engaged in surveillance practices upon both their own citizens and foreign governments that are empirically similar to those labeled "cybercrime" where they are perpetrated by other actors. These practices, then, under these circumstances, occupy both a legal and a sociopolitical gray zone where their location within legal/illegal and licit/illicit schema are potentially more fluid and difficult to determine. Furthermore, the social categorization of these acts as illicit is not universal. There is evidence that cybercriminals enjoy nonpariah, even heroic, status in some regional contexts such as that of Nigeria in West Africa. In Nigeria cybercrime directed at victims from the global North is commonly viewed as a justifiable response to the country's troubled relationships with former colonial powers and more recent histories of environmental exploitation by rapacious Northern corporations. These macrohistorical circumstances have coincided with local traditions of technological literacy that found few openings within the legitimate regional economy and which led to the early pioneering of fax and e-mail scams in Nigeria. Here the social acceptance of 419 cybercrime scams (in which an individual is tricked into advancing money to a stranger) is evident through their being celebrated in pop cultural discourses such as the hit song "I Go Chop Your Dollar" (Glenny, 2008: 207–210). Cybercrime, then, in Nigeria appears to be both illegal and licit at the same time. The legal and social categorization of all potentially criminalized acts, therefore, is not absolute but is more relational in nature and a product of their spatial (and temporal) contexts rather than any inherent quality of these acts themselves.

ENVIRONMENTAL CRIME

Organized criminal groups also undoubtedly profit from the illegal exploitation of environmental resources including minerals, precious metals and stones, and plants and animals, and from the abuse of the environment through the illegal transport and dumping of hazardous substances and waste of various kinds (Walters, 2010). Globally, aspects of these illegal environmental economies have been estimated as constituting extensive revenue streams for some criminal groups: "For example, a recent report by the World Wildlife Fund (WWF) estimated that the value of illegally poached game, and the sale of products such as ivory and rhino horn, was USD 19 billion" (Interpol, 2014: 7; see also United Nations Office on Drugs and Crime, n.d.). These extensive trades tend to emanate now predominantly, but not exclusively, from the countries of the global South where criminal groups seek to take advantage of asymmetries in environmental regulation and enforcement.

The direct and indirect environmental consequences of illegal economies such as those of narcotics production, which include deforestation, habitat depletion, and pollution of ecosystems, have also been recorded anecdotally (Laville, 2008; Parry, 2008) and through some pioneering critical academic studies (Álvarez, 2002; Dourojeanni, 1992). It is important, however, not to consume official discourses of environmental crime uncritically. The foci, within these discourses, on the involvement of organized criminal groups tend to obscure the historical significance of violent ecologically and socially harmful environmental exploitation and degradation in early state formation (Hobbs and Antonopoulos, 2013: 31) and the now far greater environmental harms generated through corporate and state practices (Walters, 2010). While not discounting the significance of illegal environmental economies, it should be recognized that in tending to isolate the roles of organized criminal groups in these trades, such discourses reproduce the fallacy of environmental crime as an exogenous threat.

MONEY LAUNDERING

"Money laundering," the passing of illicit finance via circuitous routes through the financial system to create the impression of licit income by obfuscating its criminal origins, is a process that underpins the economies of organized crime. The anonymity and opacity of the now globally extensive, offshore economy (see Chapter 5) offers considerable benefits to criminal entrepreneurs seeking to launder their revenues. Measuring the extent of laundered finance, at any scale, is a challenge given that its intention is to disguise the illicit origins of such movements. Estimates of the extent of laundered finance generally place it at somewhere between 2 and 5% of global GDP (Levi, 2002: 879; Truman and Reuter, 2004). The United Nations Office on Drugs and Crime (2011: 5) estimated that US$1.6 trillion, or roughly 70%, of criminal proceeds were laundered globally in 2009, which is equivalent to 2.7% of global GDP. Although this estimate came with many caveats and qualifications, it was broadly consistent with a range of estimates that preceded it. Whatever the precise figure, money laundering and other illicit financial mobilities are certainly significant in their global extent.

Furthermore, there is evidence that these illicit financial flows play important roles in the maintenance of the global financial system, particularly in times of recession when licit streams of finance to institutions drop (Syal, 2009: n.p.; Vulliamy, 2012: 28). Money laundering, like the offshoring of finance generally, is not just a "paper" or digital exercise, however. It is one that can have significant material impacts within vulnerable regions and has been implicated in national economic crises such as the 2001 crash of the Republic of South Africa rand (Nordstrom, 2007: 168).

ORGANIZED CRIME AND THE STATE

Organized criminal markets, like all economic activities, are not aspatial. As alluded to above, they are richly patterned and highly uneven across global space and within spaces at other scales. Despite this reality, though, the spatial has rarely been a substantive dimension of analysis within the literatures of organized crime (Hall, 2012). While it is often noted or alluded to in these literatures, which have originated primarily in disciplines such as anthropology, criminology, political science, and sociology, it has rarely been pursued to any substantive degree (Carter, 1997; Glenny, 2008; Hignett, 2005; Siegal, 2003). Van Dijk's (2007) work, which is discussed in more detail in Chapter 3, is a rare overtly spatialized reading of global patterns of organized crime. Van Dijk's Composite Organized Crime Index (COCI) offers a composite measure of perceptions of organized crime that are mapped by country. This analysis highlights the significance of weak rule of law and absences of various dimensions of good governance as closely related to regional concentrations of organized crime. This echoes the findings of other studies (Abraham and van Schendel, 2005; Aning, 2007; Castells, 2000; Hall, 2010b; Wright, 2006). However, it is worth noting that elsewhere high levels of organized crime have been observed in strongly governed contexts (Glenny, 2008; Hill, 2005; Levi, 2014: 11; Scott-Clark and Levy, 2008; Varese, 2011).

Elsewhere van Dijk cites the potential of the "resource curse" to generate rampant corruption in countries that are richly endowed with natural resources but suffer from weak institutional capacity and the presence of various internal and external predatory actors. Such resource curses are most commonly associated with natural resources such as oil and diamonds, whose exploitation has generated instability, conflict, and aggressive rent-seeking behavior, undermining effective governance by the state, in places such as Angola, Nigeria, Venezuela, and the Democratic Republic of the Congo (van Dijk, 2007: 49), but it can also be associated with resources derived from illicit trades in drugs, which undoubtedly influence the broad contours of the international criminal landscape that were discussed above.

Post-Soviet space, reflecting van Dijk's analysis, has been identified elsewhere as a key terrain in the recent history of organized crime (Bhattacharyya, 2005: 63; Glenny, 2008). It is thought that during the 1990s roughly 40% of Russia's economy was controlled by organized criminal groups (Dunn, 1997; Castells, 2000: 184; Frisby, 1998: 34) who exploited the "hiatus between the demise of state power and the development of a true civil society" (Wright, 2006: 155). While the power of organized crime in Russia has declined since then, Russian organized criminal groups are now among the most internationalized of criminal organizations, spurred in part by instability within Russia (Galeotti, 2005b; Siegal, 2003). Although these groups are undoubtedly parasitic, some authors note their role in facilitating the transition to capitalism across post-Soviet space (Aas, 2007; Castells, 2000; Stephenson, 2015; Wright, 2006).

High levels of organized crime are also apparent in other parts of Eastern and Central Europe (Hignett, 2005; Wright, 2006: 151–153). In Georgia, for example, "thieves-in-law" (gangsters) have controlled between 30 and 60% of sectors such as finance, construction, and gambling and also enjoy considerable "normative influence" in the context of poverty and an absence of strong public attachment to the state (Slade, 2007: 177–179). These normative, cultural impacts, particularly on young people, of gangsterism, corruption, and its associated conspicuous consumption, have been noted in both post-Soviet contexts and beyond (Bayart, Ellis, and Hibou, 1999; Slade, 2007; Stephenson, 2015).

The case of Afghanistan reveals the complex relationship between organized crime and state outcomes. While the power vacuum since the 1992 Soviet withdrawal allowed drug production and other illicit trade to flourish, with between 30 and 60% of Afghanistan's gross national product (GNP) equivalent stemming from opium poppy production, the drugs economy is one of the few mechanisms through which forms of governance and order penetrate its peripheral regions. Goodhand (2009: 22) argues, for example, that "whereas other markets in Afghanistan are extremely fragmented, the drugs economy represents the nearest thing to a national market, involving multiethnic networks and strong north–south integration."

Elsewhere it has been noted that high revenues from drug production have made important contributions to employment and national income in certain nations (Bagley, 2005: 36). Colombian cocaine export revenues, for example, have totalled US$3.5 billion per annum (1999), equaling income from oil exports and exceeding that from coffee (Bagley, 2005: 38; see also Castells, 2000; Kenney, 2007). Extensive drug production has also become established within deindustrialized regions of the global North. Cannabis production generates 5% of British Columbia's GDP and provides employment for approximately 100,000 people, more than are employed within the region's traditional industries (timber, mining, oil, and gas) (Glenny, 2008: 245–255). Certainly, the significance of drug money to certain regional economies and the extent to which this illegal market can embed itself within regions poses serious challenges to antidrug policy.

> Drug economies may be scary and violent, creating societies that cannot guarantee the safety of ordinary people or the autonomy of state institutions, but they are resilient. . . . Dislodging the hold of the drug trade also threatens to undermine the livelihoods of a host of ordinary people who have few options other than to participate in the major economic activity at hand. Once drug money has seeped into all areas of life, hitting the trade in drugs hits everyone. (Bhattacharyya, 2005: 118–119)

There is a lack of clear evidence and consensus within the literature concerning the contributions of organized criminal economies to state-making and development trajectories, particularly in parts of the global South where legitimate alternatives may be lacking (Bhattacharyya, 2005: 118; Midgley et al., 2014: 23, 26). While criminal markets undoubtedly make, in some cases considerable, contributions to regional economies (van Dijk, 2007: 50; Goodhand, 2009; Lee, 2008: 345; Nordstrom, 2007), others have argued, citing cases of violence, undermining governance capacity, state capture, and the diversion of resources from the legitimate economy, that the net economic and political impacts of organized crime are inevitably and overwhelmingly negative (Bhattacharyya, 2005: 96; Glenny, 2008: 291; Jesperson, 2017). Van Dijk is categorical, stating that

> the overall net effect of organized crime on the economy through the paths of weakened governance and rampant corruption, is strongly negative. . . . A national growth strategy based on tolerating organized crime, is not just immoral but economically self defeating. Under such policies no country will ever achieve sustainable development. (2007: 54)

Where criminal economies prevail within their regional contexts, they can certainly deliver the prerequisites for sustainable development, namely, financial capital, entrepreneurialism, governance and organizational capacity within the economic realm, the skilling of the labor force, and the provision of infrastructure. However, the presence of these inputs, the capacity to develop sustainable, licit economic trajectories, does not means these potentials will necessarily be realized, certainly in senses imagined in prevailing Western sustainability discourses. It is possible to imagine barriers to such models of sustainable development in such cases. Internally, there is likely to be little willingness to legitimize the economy. Criminal economies—for example, those centered around the production of narcotics—can be very lucrative relative to licit economic alternatives (Wainwright, 2016). Despite the risks involved, it is unlikely that criminal entrepreneurs will willingly forego these greater rewards. Furthermore, the identities of key actors in these markets are often complex, spanning multiple licit and illicit sectors and activities in ways that produce interdependencies between them (Goodhand, 2009: 18–19). These markets, then, are embedded in the economic, political, and social structures of their regions in ways that make their removal or transformation problematic. Furthermore, such economies are difficult to penetrate by external agents seeking to influence their development trajectories. Local institutional capacity is also often corrupted and hence acts as a further drag on change. The literature, then, records only very few examples of formerly criminal economies that have been "legitimized" and continue to develop in conventionally sustainable ways through, for example, alternative development strategies (Clemens, 2008; Farrell, 1998; Felbab-Brown, 2014: 41, 45; Lupu, 2004; Rozo, Gonzalez, Morales, and Soares, 2015). Even where this has been recorded, though, there is compelling evidence of illicit production merely being displaced to other regions rather than being eradicated entirely (Felbab-Brown, 2014: 41; McCoy, 2004).

Successful external interventions into criminal markets are more likely to be disruptive and thus damaging to the future growth prospects of these regions (Cohen, 2009). Indeed, in some cases, counternarcotics operations have been identified as potential threats to regional growth (Goodhand, 2009: 23), so significant are illegal economies to such regions. The literature, though, is replete with examples of extensive, long-standing criminal markets (Bhattacharyya, 2005; Castells, 2000; Costa and Schulmeister, 2007; Galeotti, 2005a). While these may not fit prevailing definitions of sustainable development, and are highly problematic in a number of ways, some authors who have explored them have asked if they deliver greater development impacts than licit alternatives (Bhattacharyya, 2005). Certainly, Goodhand's (2009: 22) analysis of illicit opium-based markets in Afghanistan recognizes their roles in the development of the country's peripheries. These subjects are returned to in more detail in Chapter 6, where the impacts of criminal mobilities on the regions through which they move are discussed.

CRIMINAL ORGANIZATION IN A GLOBAL ECONOMY

Changes to the economic contexts within which organized criminal groups operate and the increasingly international terrains traversed by criminal networks have undoubtedly affected the bases upon which these groups are organized (Hobbs, 2013; Scalia, 2010). Territory and ethnicity have become less significant foundations in the organization of criminal activities (Galeotti, 2005a, 2005b; Hobbs, 1998, 2001, 2004, 2013; Kleemans and van de Bunt, 1999; Serious and Organised Crime Agency, 2006). While geographical differences between criminal groups might be declining (Nordstrom, 2007), place has certainly not become irrelevant to the patterning of organized criminal markets. There is a wealth of evidence from the literature to show that there remain rich local geographies of criminal organizations and their activities, despite the hegemonies of global discourses and the apparent leveling effects of globalization. However, while empirically revealing these spatialities, these literatures have perhaps underplayed their significance conceptually. Indeed, some have gone so far as arguing that these differences are superficial and have little tangible effect on the shape of the organized crime landscape (Albini et al., 1997: 154). Here, though, we can look to critical geographical literatures of the globalization of the licit economy that have noticed the endurance and transformation of local distinctiveness, rather than its erasure, under powerful processes of globalization and incorporated them within their articulation of the contemporary global economy (Dicken, 2011; Sheppard, 2002). Questions of criminal organization under globalization are returned to in more detail in Chapter 4.

Organized crime continues to be influenced by the cultural and economic characteristics of its regions (Calderoni, 2011; Weinstein, 2008; Wright, 2006: 5). Thus, at the most basic level, there is clear evidence of enduring differences in the organization, activities, and cultures of criminal groups from different nations and regions (Aning, 2007; Davidson, 1997; Hastings, 2009; Hignett, 2005; Hobbs, 2004; Hobbs and Dunnighan, 1998; Levi, 2002; Lintner, 2005; Lombardo, 1997; Scalia, 2010; Varese, 2011). The internationalization of criminal organizations does not make then suddenly "free floating" or ontologically ungrounded from the spaces in which they operate. Despite the increasingly international orientation of organized criminal groups, rootedness within specific locales remains important to their identities and activities (Castells, 2000: 183; Grascia, 2004;

Hill, 2005; Kleemans and van de Bunt, 1999; Madsen, 2009: 4–5; Scalia, 2010; Varese, 2011). Paoli (2005: 23), for example, notes the ways in which preconditions for recruitment to the Sicilian Cosa Nostra that demand not only that members were born in Sicily but that families do not settle outside of Sicily, has hampered their involvement in international markets. Varese's (2011) extensive empirical analysis of mafia migrations suggests that criminal capital is a difficult commodity to mobilize and is derived from the groundedness of criminal organizations in specific spaces. However, there has been little explicit reflection on the significance of these geographical differences and their influence on the nature and impacts of organized crime, something this book moves on to address across its remaining chapters.

Summary

This chapter has demonstrated that organized criminal markets of many kinds form extensive elements of the contemporary global economy. It has attempted to explicitly outline some of the geographies of these economies, something that has been, at best, implicit in most of the extant literatures devoted to organized crime, regardless of the disciplines from which they derive. It has also attempted to foreshadow to an extent a number of the themes that are explored in more detail in subsequent chapters. At many points during this chapter notes of caution were sounded over the validity of sources of data and evidence that is available for the study of organized crime. It is to this vital question, and to the questions of researching and measuring organized crime more generally, that the next chapter turns.

CHAPTER 3

MEASURING AND RESEARCHING ORGANIZED CRIME

INTRODUCTION

Given the prominence attached to claims about the extent of organized criminal activity across a variety of terrains, it is perhaps surprizing to find little critical discussion of approaches to quantitatively measuring and estimating its scale, save for a small literature that begins to critically explore technical, conceptual, and, to an extent, political terrains (Hobbs and Antonopoulos, 2014; Holmes, 2016; Levi, 2014; Midgley et al., 2014: 18). Thus, commentators are able to plausibly argue that measuring the nature, extent, and scale of organized crime and trends therein is an extremely difficult endeavor (Holmes, 2016: 42), that data are often missing or are unreliable and should be treated with caution (Inkster and Comolli, 2012: 15–16; Levi, 2014: 8; Midgely et al., 2014: 18; Watt and Zepeda, 2012: 4–5), but that this should not stand in the way of efforts to expand our understanding in these regards (Calderoni, 2011: 42; Hudson, 2014: 781). Indeed, there are powerful policy drivers to come up with statistical estimates of the extent of organized crime in a variety of contexts (Sharman, 2011: 18–20).

Similarly, while the academic literatures on organized crime have produced more diverse, nuanced, and often qualitative knowledges of its subject, they have tended to reserve relatively little space for the critical, reflective, or instructive discussion of the methods employed to underpin their analysis, despite the considerable methodological, ethical, legal, and safety implications associated with researching illicit and illegal activities generally and particularly those associated with high levels of violence such as organized crime (Brooks, 2014; Castells, 2000; Garrett, 2014; Renzetti and Lee, 1993). Methodological discussions in the majority of research articles and books on organized crime are somewhat bloodless affairs. Scholars in this area have generally not taken the opportunity to take the reader with them into the field. While the field of research into organized crime presents many barriers for those researchers encountering it, there is only a single substantive publication to date whose explicit focus is instructing researchers about the ways in which they might negotiate these barriers (Hobbs and Antonopoulos, 2014), although there is a somewhat more expansive literature on researching illicit, illegal, and dangerous subjects more generally (Brooks, 2014; Lunn, 2014).

This chapter turns a critical eye on questions of researching organized criminal markets in the broadest sense. It is concerned with official sources of data that attempt to measure the extent of these markets within nations, regions, or the entire globe; reports within journalistic, biographical, and "true crime" genres; and also the approaches employed by academic researchers. The chapter begins by looking at various official counts and measures that have been developed in the examination of organized crime before moving on to consider transnational perceptions and experience surveys and the construction of various indexes of organized crime. It then considers the notions of the costs and harms of organized crime and the ways in which they inform the collection of data. The chapter next explores the potentials of the relatively recent emergent approaches of threat or risk assessments. Subsequent sections consider examples of serious and insightful investigative journalism that have done much in recent years to open up some of the more violent and elusive worlds of organized crime to external scrutiny. By contrast the chapter then outlines the flawed, sensationalized renditions of organized crime typical of the "true crime" genre and gangster autobiographies. The largely as yet untapped potentials of social media as sources of data for students of organized crime are then considered before the chapter moves on to examine academic research approaches, including historical and ethnographic accounts and critical geopolitics, that have either been applied to the study of organized crime or have the potential to enhance extant approaches. The chapter concludes with some thoughts on research and methodological priorities for scholars of organized crime.

The chapter does not attempt to offer a "how to" guide. These issues are more directly addressed elsewhere (see, e.g., Hobbs and Antonopoulos, 2014; Holmes, 2016; Levi, 2014). Rather, it offers a critical reflection on questions of researching and measuring organized crime and the relations between the knowledges these endeavors produce and institutional, legislative, and policy responses to organized criminal activities. In doing so, it considers questions such as What is measured/researched? How are these measured/researched? In what ways are these knowledges presented and utilized and by whom?

OFFICIAL MEASURES OF ORGANIZED CRIME

The ambiguities and debates around what exactly constitutes organized crime and the failure to arrive at a universally agreed-upon definition of it are well known and were laid out in some detail in the opening chapter of this book (see also Calderoni, 2011; Hobbs and Antonopoulos, 2014; Holmes, 2016: 29; Levi, 2002; Madsen, 2009; Wright, 2006: 2–14). While this is a distracting conceptual debate, practically it makes measuring and researching organized crime very difficult. Simply argued, it is very difficult to measure something that has not been, and appears unlikely to be, clearly and universally defined. "Organized crime" refers to an incredibly diverse terrain. It might be asked how practical and meaningful is it to attempt to measure and speak of activities as diverse and dispersed as cybercrime, of various types and degrees of sophistication originating from places such as Nigeria, Brazil, and Eastern Europe, among many others (Glenny, 2008, 2011; Lusthaus, 2013; Neal, 2010); industrial-scale drug production and trafficking through the Americas (Brophy, 2008; Bunker, 2011b; Vulliamy, 2010; Watt and Zepeda, 2012); and, for example, relatively small-scale localized trading in stolen or smuggled goods such as cigarettes and clothing among marginalized urban communities in the cities of the global North (Hobbs, 2013; Hornsby and Hobbs, 2007). However, when statistics speak of organized crime, they are speaking of all of these activities and a whole lot more besides. Rhetorically, then, the term "organized crime" elides criminal activities for which there is little connection or commonality empirically. This is an important part of the wider, long-standing politicization of the concept of organized crime as an existential and security threat to nations (Woodiwiss and Hobbs, 2009).

The boundaries of organized criminal activities, in terms of their relations to other volume crime activities; other serious crimes, most notably terrorism (Madsen, 2009: 75); and licit economic realms are ambiguous, fluid, and transitional rather than anything like sharply defined and static. This book, at many points, reflecting the prevailing view of critical literatures, argues that organized criminal activities overlap and are deeply interrelated with a host of seemingly licit economic realms and their associated activities (Aas, 2007: 125; Bhattacharyya, 2005: 63–64; Castells, 2000; Hudson, 2014; Nordstrom, 2007, 2009; Weinstein, 2008; Wilson, 2009; Wright, 2006: 52). Untangling at these edges what is licit and productive and what is illicit organized criminal activity is highly problematic empirically, practically, and ontologically.

Organized criminal activities might be thought of as processes rather than as autonomous acts. This is most clearly demonstrated through the various stages involved in the production, distribution, and consumption of illicit commodities and the need for organized crime groups to launder their profits and, in doing so, distancing them from their criminal origins (Levi, 2002: 879; Truman and Reuter, 2004; Sharman, 2011; United Nations Office on Drugs and Crime, 2011: 5). It is through these processes that ambiguities and overlaps with the licit often emerge. For example, it has been widely observed that laundered criminal finance has been invested in real estate and urban development projects (Glenny, 2008; Weinstein, 2008), thus entering sectors of the economy that are, at the very least, not overwhelmingly populated by criminal actors, although they, and those actors who support their presence, are undoubtedly present to some extent. This process view of organized crime raises a number of issues for those engaged in researching and particularly in measuring it. Practically, if the professionals engaged in the laundering process have adequate competency, it will be extremely difficult for researchers and investigators to observe the movement of laundered criminal finance within these contexts and separate it off from finance originating from licit sources. Also, if organized crimes are processes rather than single autonomous acts, they present those charged with measuring them with the challenge of what point along these processes do they make their recordings (Holmes, 2016: 40). Recordings made at the point of sale within illicit consumption markets, for example, would give very different quantitative value measurements to those made at various points of production and refinement or within the laundering process, given the costs that these processes incur and price escalation along the supply chain (Wainwright, 2016: 239). Furthermore, the bald statistics obtained at any single point within the production, distribution, consumption, and laundering process would say nothing of the complexities of the distribution of economic costs and impacts across the often international spaces associated with the various stages of these processes. Indeed, if statistics are being obtained on a national basis, they might inherently exclude costs and impacts associated with stages of production and distribution occurring beyond immediate national boundaries. Given the multiple stages involved in these processes, it is highly unlikely that all stages would equally be available for scrutiny, producing, most likely, incomplete and uneven pictures of these criminal economic processes. Finally, it has been argued that the economic revenues and investments associated with organized criminal activity can potentially produce developmentally positive as well as negative outcomes, although this is a highly contested point. An example of the former would be the increases in wages, credit, and consumption in the rural economy observed in the regions of Afghanistan involved in its economy of opium production and trafficking (Goodhand, 2009: 22; see also Bhattacharyya, 2005: 118; Hall, 2013: 372–373; Hudson, 2014: 790–791). Typically, though, organized criminal activities are measured through the negatives of costs, harms, and threats (Holmes, 2016; Levi, 2014). The facts of potentially positive outcomes and impacts from these processes have gained no purchase within the instruments of measurement of organized criminal economies, giving, arguably, skewed impressions of the effects of these markets. However, while such debates have been aired within some academic contexts, there has been little room for such complexities within the official statistical discourses of organized crime. The object of measurement here, then, organized crime, is beset by this series of uncertainties, unknowns, and ambiguities.

Setting aside debates about what exactly is being measured for a moment, organized crime has been measured through a variety of different metrics. The four main metrics employed are various measures of extent, such as counts of the numbers of criminals or criminal groups estimated to be operating within particular markets and/or jurisdictions or measures of the revenues derived from various organized criminal activities (see, e.g., Serious and Organised Crime Agency, 2006); these are typically based on police intelligence, seizure and case data, and various measures of the costs of organized crime, These are economic expressions of the resources required to tackle and manage organized crime, sometimes but not always including some indirect costs such as those of treating the effects of drug abuse (Levi, 2014: 70) or the harms generated by organized crime that exceed the purely economic measures of cost and might, for example, include the social and environmental costs of organized crime and which often draw on victimization, perception, or experience surveys; and finally expressions of the threats posed by organized crime (Hobbs and Antonopoulos, 2014; Holmes, 2016; Levi, 2014; Sproat, 2012). These different approaches produce very different economic measures of organized crime. For example, it has been estimated that the value of the cross-national trafficking of heroin along the Balkan route into Europe is worth between US$6.7 and US$10 billion per annum. However, if the value of production and consumption of this traffic is included, then its value rises to US$28 billion per annum. This figure, though, is many times lower than the one produced if the health and enforcement costs associated with it are also included (United Nations Office on Drugs and Crime, 2015a; cited in Berlusconi, Aziani, and Giommoni, 2017: 2).

Sproat (2012: 314–316) provides a critical account of the policing of organized crime in the United Kingdom. Within this account he deconstructs the British government's discursive framing of organized crime as an extensive and serious threat to the nation. Central to this framing is the deployment of officially sanctioned and shared estimates of the extent of organized crime in the United Kingdom. This discussion includes details from a variety of quantitative assessments of organized crime in the United Kingdom. These sources include claims such as there were 25,000 to 30,000 criminals engaged in organized crime in the United Kingdom in 2009, activities that generated illicit turnover of £15 billion per year (2007), and economic and social costs to the country of between £20 and £40 billion (Sproat, 2012: 314, 316; see also Levi, 2014: 9, who reports U.K. Home Office estimates of the economic and social costs of organized crime in the United Kingdom as £24 billion in 2013). On the basis of data such as these, the United Kingdom's 2008 National Security Strategy was able to argue that organized crime represented a "serious and fast moving threat" (Sproat, 2012: 314) comparable to that of terrorism. These were views of organized crime in the United Kingdom shared by a number of institutions including the U.K. police force's Serious and Organised Crime Agency, HM Treasury, the Home Office, and the Foreign and Commonwealth Office (Sproat, 2012: 316).

While appearing plausible and being widely circulated and consumed within political, law enforcement, and media circles, such official estimates have drawn a wide range of criticisms from academic scholars. Data such as these have been criticized, for example, for their methodological flaws, vagueness, and lack of clarity (Hobbs and Antonopoulos, 2014: 3); the lack of internationally standardized reporting procedures (although the European Union has begun efforts to address this issue); the very different reporting cultures and practices that exist within and between different countries (Holmes, 2016: 29–31, 38–39, 43; Levi, 2014: 9); the variable competencies of agents whose work underpins official sources of data and the unreliability of those reporting official data (Hobbs and Antonopoulos, 2014: 3; Holmes, 2016: 30); and the inherent guesswork that is a part of estimating activities that are inevitably clandestine by nature (Holmes, 2016: 31). While these are largely technical issues that are likely to render official estimates of organized crime markets, their costs, harms, and threats objectively inaccurate, some have recognized a systematic exaggeration of the threat posed by organized crime. These exaggerated threats, it has been argued, have been deployed through official discourses for two reasons. The first is what has been termed "governing through crime," the political exaggeration of threat as a justification for the extension of criminalizing, punitive, repressive, or intrusive legislative change and the restriction of fundamental rights for citizens (Simon, 2007). This phenomenon has been observed in a variety of international contexts (Lajous, 2014). The second represents the exaggeration of threat by law enforcement agencies as part of their attempts to secure resources within increasingly competitive national public funding contexts. This has been a widely observed and commented upon international phenomenon (Abraham and van Schendel, 2005: 2; Hobbs, 1998: 139; Holmes, 2016: 40; Kleemans, 2008: 6; Lee, 2008: 336; Levi, 2002: 895; Woodiwiss and Hobbs, 2009; Wright, 2006: 23).

With regard to the official U.K. estimates of organized crime referred to above, Sproat reflects this skepticism and concludes by "wonder[ing] about the validity of the official claim that organized crime in the UK constitutes any sort of tangible—as opposed to theoretical—threat to the UK's national security—never mind a serious one" (2012: 328). The structural characteristics of organized crime groups, which tend, predominantly, to be small, flexible, networked, often ephemeral units, and the absence of centralized organized crime conspiracies (Galeotti, 2005a; Glenny, 2008; Hobbs, 2001,

2004; Hudson, 2014; Kleemans and van de Bunt, 1999; Scalia, 2010; Stephenson, 2015; Varese, 2011; Wright, 2006) (see Chapter 4) militates somewhat against their constituting a serious threat to security on a national level in the majority of instances. Thus, Levi is able to argue that "few individual elements of organised crime constitute a national-security threat; the extent to which both the criminal actors and the impact of their crimes *in their totality* [original emphasis] represent a threat to national security also remains contested" (2014: 12). Despite all this, statistical measures of organized crime continue to be deployed by official bodies, seemingly in efforts to amplify the appearance of their threat.

In summary, then, where the deployment of the term "organized crime" has been criticized as a political act (Hobbs, 1998, 2013; Levi, 2009; Woodiwiss and Hobbs, 2009), so too, logically, should attempts to speak of it through official statistics be seen as potentially similar political utterances. There is a danger, as they are currently deployed, for measures of organized criminal activity to be as much exercises in expressing a politicized rendition of the market economy, founded on the myth that it is composed of a series of discrete realms, as it is an exercise in recording an external empirical reality. Thus, it has been argued that "official data are the result of law enforcement activity, which in turn is the result of resource restrictions, the competency of agents, organizational priorities, and wider political priorities" (Hobbs and Antonopoulos, 2014: 3).

Given the range of technical, conceptual, and political concerns outlined above, this begs the question of how should critical scholars of organized crime approach and potentially use official sources of data about their subject. Should they, for example, be dismissed as little more than the political exercises that Hobbs and Antonopoulos discuss above or can they be a resource deployed within critical narratives in ways that challenge politicized renditions of organized crime? Certainly, researchers should not automatically reject official sources and especially archives. As Hobbs and Antonopoulos (2014: 2) point out, there have been many classic studies of organized crime that have drawn, at least in part, on official case file archives, if not necessarily official statistical data. They list Blok (1974), Catanzaro (1992), Gambetta (1993), and Hess (1973), to which we might add Varese (2011) and many more specific studies (e.g., see Schoenmakers, Bremmers, and Kleemans, 2013: 325). These studies were primarily focused, however, on the exploration of the internal structural characteristics of criminal groups at a typically local scale rather than assessments of their externally directed and extensive costs, harms, or threats. Notwithstanding the editing processes undertaken by the authors of these archives and the restrictions placed on them by gatekeepers (Hobbs and Antonopoulos, 2014: 3), these sources do offer potentially valuable resources to those scholars prepared to approach them critically and perhaps apply them to their own research agendas rather than to simply replicate the agendas of the authors of official archival sources.

It is worth also, then, reflecting on the potentials for official sources of statistical data relating to organized crime to be available as resources for those engaged in the development of critical narratives. It is true that these statistical data have tended to be most readily appropriated by those whose aim is to realize politicized renditions of organized crime that sketch it as an extensive, external, existential threat to a pristine mainstream (Woodiwiss and Hobbs, 2009). However, while regrettable, given that this view often does more to cloud than illuminate understandings of the roles and places of the illicit in the contemporary world, this does illustrate that the interpretation of such data is, in part at least, a product of the broader narratives into which they are inserted. In Chapter 2 of this book, for example, and elsewhere (Hall, 2013; Hudson, 2014), these data have been used to rather different ends. Here they are inserted into narratives whose aims are more to highlight the idea that while organized crime is an extensive element of the contemporary economic landscape, it is neither external to the mainstream nor an existential threat, but rather a part of the normal relations and processes of this global landscape of exchange. Here these data are deployed to support these claims. While it remains vital to foreground the flaws and limitations of the data that are available to describe organized crime in various ways, in these alternative narrative contexts they become part of a very different set of stories to those that have otherwise prevailed.

PERCEPTION AND EXPERIENCE SURVEYS OF ORGANIZED CRIME

Perception and experience surveys, directed at either citizens or businesses, are widely used methods employed to assess the scale and costs of organized crime. There are a number of prominent official assessments of organized crime that rely on these approaches including the annual World Economic Forum's *Global Competitiveness*

Report (Schwab and Sala-i-Martín, 2015) and the *Business Environment and Economic Performance Survey,* an experiential survey produced by the World Bank (Holmes, 2016: 33). There are concerns about validity of perception surveys as they are an indirect measurement of, in this case, organized crime. It is argued that they do not reflect the reality of organized crime but rather publics' and business communities' perceptions of this concept, which might be objectively inaccurate. There is evidence also that perceptions might lag behind change in, in this case, levels of organized crime within a region (Midgley et al., 2014: 18). Indeed, perception surveys have produced some curious responses, in some cases highlighting organized crime as an apparently serious issue in places where there is little empirical evidence of any kind to sustain this perception. However, this criticism has been countered by arguments around the actual elusiveness of any objective measure of organized crime and the view that perceptions, rather than merely reflecting a prior reality to varying degrees of accuracy or inaccuracy, actively constitute reality (Crang, 1999: 60; Holmes, 2016: 32). Levi (2014: 8), for example, has argued that measuring the perceptions of organized crime is a valid exercise as these perceptions themselves constitute one of the social harms associated with organized crime. However, it is difficult to avoid the limitations of public perception and business experience surveys conducted in areas where organized crime is deeply embedded into the local economic and social environment and where respondents might be justifiably reluctant to report their experiences of organized crime to the authorities for fear of reprisals from violent criminal actors (Holmes, 2016: 34) or where conditions on the ground preclude the collection of such data. Whatever the circumstances, surveys of this nature should be interpreted with reference to the conditions prevailing within their regional contexts: "Simply measuring public opinion without understanding the context of fear, intimidation and the role of civic tradition may not provide any clear indication as to the possibility of [community] mobilisation [against organized criminal groups]" (Midgley et al., 2014: 29).

ORGANIZED CRIME INDEXES

Critical scholars have on many occasions drawn on official sources of data in their mappings of organized crime through the construction of indices that attempt to measure the presence of and extent of organized crime and its variations across space. Examples include van Dijk's (2007) global Composite Organized Crime Index (COCI) (Holmes, 2016: 36); Calderoni's (2011) Mafia Index, which attempts to measure the presence of crime groups across Italian regions; and Lavezzi's (2008) analysis of the economic structure of Italian regions in an attempt to uncover the presence of factors that are associated with Mafia presence, although there is also a wider tradition of such index building within the official and criminological literatures of organized crime (Calderoni, 2011: 42–53).

Van Dijk's COCI (Figure 3.1) combines a number of measures of the perception of organized crime with the objective measure of the rates of unsolved homicides within individual nations. The latter is taken as a measure of the extent of instrumental violence that is associated with organized crime markets. In terms of perception data, van Dijk drew on results from World Economic Forum victimization surveys conducted among business executives that explored their experiences of racketeering and the collection of protection money by organized crime groups. These surveys have been conducted since 1997 and across 102 countries and each year included a sample size of 500 or more. In the calculation of the COCI, van Dijk drew on the surveys from 1997 to 2003. The COCI also included data on investment risk associated with the prevalence of organized crime across more than 150 countries, which was undertaken by the London risk consultancy Merchant International Group. To these measures the COCI included data on levels of corruption of public officials and the extents of the informal economy and money laundering in different nations. The corruption data came from data sets from the World Bank Institute, while those on the informal economy and money laundering came from World Economic Forum surveys of business executives (van Dijk, 2007: 40–42). Of this index van Dijk argues:

> An important strategic advantage of the resulting Composite Organized Crime Index is the incorporation of at least one objective measure of organized crime activity, the rate of unsolved homicides according to official administrations. Scores on this composite index cannot be dismissed by governments as being based on "just perceptions." (2007: 42)

Much debate within the literature centers on methodological criticism and the articulation of technical flaws or limitations with the methodologies and/or data sources underpinning specific indices.

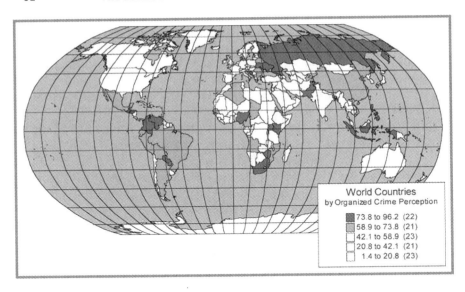

FIGURE 3.1. Global mapping of Composite Organized Crime Index (COCI). From van Dijk (2007: 44). Reprinted by permission of Springer.

Holmes, for example, cautions against uncritical acceptance of van Dijk's (2007) findings given their heavy reliance on perception data (2016: 36; see also Calderoni, 2011). While it is improbable that we will arrive at an empirically perfect index of any kind—indeed, we would not know we had arrived at such a place should this ever occur—such models do provide a selection of data-based lens through which to examine organized crime and frequently highlight broad factors that equate or correlate to varying degrees with the presence of criminal groups and markets within regions. As such, they may point broadly to potential policy directions. Such indices are also useful in challenging stereotypical perceptions of the geographies of organized criminal activity that suggest they are restricted to a small number of regions (Hobbs and Antonopoulos, 2014: 5). One of Calderoni's key findings, for instance, was that while Mafia distributions in Italy remain strongly reflective of their territories of origin, these groups are also discernible, in some cases to significant degrees, in central and northern provinces; "this confirms that mafias should not be regarded as typically Southern Italian phenomena, but rather as a national problem" (2011: 41). This finding was a little at odds, though, with the conclusions of Varese's (2011) study, based on case studies of specific Mafia groups, which stressed the relative immobility of organized crime groups.

THREAT AND RISK ASSESSMENTS

A method for assessing organized crime that has gained increasing traction within law enforcement circles since the late 1990s is that of threat (or risk) assessment. Threat/risk assessment approaches are less concerned with the production of decontextualized counts of various indicators of organized crime, and more concerned with structural, environmental assessment of the threat of organized crime in particular contexts (Holmes, 2016: 37–38). These assessments offer a more nuanced approach to measuring organized crime in that they accommodate scrutiny of the market (licit and illicit) and institutional contexts within which organized crime operates, as well as the nature and extent of individual or collective criminal actors. These assessments are more in line with the contextualized readings of organized crime that tend to typify academic accounts. Whereas previously there appeared to be little point of contact between official assessments of organized crime and academic concerns, the growing employment of threat/risk assessment methods suggest an emergent common ground between these two perspectives. Certainly threat/risk assessments, in speaking directly of the contexts within which organized crime operates, move discussion of tackling organized crime beyond questions solely concerned with enhancing the effectiveness of law enforcement strategies. Discussing Vander Beken and colleagues (2005), Holmes suggests that threat/risk assessments point toward "a structural approach; authorities should attempt to alter market conditions so that OC [organized crime] is less able and tempted to take advantage of change" (2016: 38).

INVESTIGATIVE JOURNALISM SOURCES

There are a range of nonacademic sources available to researchers into organized crime. These include investigative journalism (Glenny, 2008, 2011; Saviano, 2008, 2016; Vulliamy, 2010), biographies and autobiographies of famous crime figures (Bowden, 2012; Escobar, 2010; Fraser and Morton, 1995; Marks, 1997), and books in the "true crime" genre (Morton, 1992, 2001, 2008; Williams, 2012). These sources, in a field in which academic engagement has been incomplete, should be acknowledged seriously, if critically and selectively, and their potential contributions to understanding recognized (Castells, 2000: 171; Dick, 2009; Hudson, 2014: 777; Watt and Zepeda, 2012: 5). In many cases the access secured by investigative journalists far exceeds that achieved, or indeed possible, by the majority of academic researchers who are fettered by ethical, health and safety considerations, and the administrative infrastructures of institutions regulating their excursions into dangerous and illicit fieldwork contexts (Brooks, 2014; Garrett, 2013, 2014). This is especially the case in regions associated with high levels of violence related to the operation of their organized crime economies. In these contexts investigative journalists have played vital roles in publicizing the problems of these regions, inaccessible to most academic researchers, but in doing so have, at times, paid a very heavy price. As Watt and Zepeda argue in the case of recent drug-related violence in Mexico:

> Some of the most insightful work on the current crisis in Mexico has been carried out by a number of outstanding investigative journalists, who often complete their work at great personal risk. For Mexico is at present among the most violent countries on Earth in which to be a reporter. In contrast, academic engagement with the topic (with some notable exceptions) has been limited. (2012: 5)

Anabel Hernàndez is a preeminent example of the type of investigative journalist discussed above, whose commentaries on the ways in which the narcotics trade has shaped, and has been shaped by, the recent political and economic landscapes of Mexico, have come at considerable personal risk. Hernàndez, who has worked on a number of Mexican newspapers, magazines, and websites since 1993, is widely recognized as a powerful voice revealing the nature and impacts of corruption and the narcotics trade and particularly the links between criminal, security, business, and political actors in Mexico. She is most well known internationally for her books *Los Señores del Narco* (2010; *Narcoland: The Mexican Drug Lords and Their Godfathers*, translated 2013b) and *México en Llamas: El Legado de Calderon* [*Mexico in Flames: The Legacy of Calderon*] (2013a). The "godfathers" of the title of the first volume refers to those leaders in the Mexican political, business, military, and police communities whose collusion in the drug trade has made is it possible, argues Hernàndez. The second volume is a damning verdict on the corruption and complicity associated with the Calderon presidency of Mexico (2006–2012). The responses of drug cartels operating in Mexico have included attacks on Hernàndez's family, multiple death threats, and dispatching the corpses of headless animals to her home (Legge, 2013).

Internationally, other credible investigative journalists who have made significant contributions to recent debates about organized crime include Misha Glenny, a former *Guardian* newspaper and BBC journalist who established an international reputation in the early 1990s for his coverage of European political change following the collapse of the Soviet Union and its ramifications across Eastern and Central Europe. He used this reputation, his contacts, and his experience to access an extensive range of sources for his 2008 book *McMafia: Crime without Frontiers*, a global survey of transnational organized crime in the post-Soviet era. The sources for this project included 300 interviews with politicians, law enforcement agents, and former and active criminal actors (2008: 395), including those involved in cannabis and cocaine trafficking, privatized violence, cybercrime, and members of Japan's Yakuza crime groups. One of the other most high-profile and well-received journalistic accounts of organized crime produced in recent years was *Gomorrah: Italy's Other Mafia* (2008) by the Italian investigative journalist Roberto Saviano. Much of the initial attention paid to the book centered on the death threat that Saviano received from his subject, the Naples crime group the Camorra, which has forced him to spend subsequent years in hiding behind constant police protection. The book drew largely on a series of covert, undercover investigations into the Camorra's activities. It contained a series of interlocked narratives detailing the daily operations and activities of the Camorra, packed with anecdotes from these close observations of a range of its actors, their associates, and victims. *Gomorrah* was subsequently dramatized as a successful feature film and later television series. *Gomorrah*'s contributions are not restricted to popular cultural circuits, however. At the time of writing (June 2017) *Gomorrah* had received almost 500 citations in academic publications. The book was also the subject of a 1-day symposium at the Extra Legal Governance Institute at the University of Oxford in 2008, which was subsequently published as a series of short papers in the academic journal *Global Crime* (Lloyd, 2009; Pizzini-Gambetta, 2009; Saviano, 2009; Varese, 2009). Saviano subsequently produced the book *ZeroZeroZero*, which investigated the contours of the global cocaine trade as a critical lens through which to scrutinize late capitalism, although critics noted the high-security conditions under which he was living at the time gave the book a more secondhand feel than its predecessor (George, 2015). Similarly, at the time of writing, the various editions of Glenny's *McMafia* had collectively received over 350 academic citations. Contributions such as these are serious and insightful perspectives that have rightly been treated with respect by academics as well as critics and the reading publics across many countries.

Beyond examples of integrity such as these, there is a regular slew of more popular, sloppily researched, or sensationalist journalistic accounts of organized crime that are of little direct use to the researcher. Even many of the better examples of the true crime genre are little more than solidly researched and written accounts that draw largely on readily available archival and media sources, enlivened at times with interviews with the associates, friends, family, victims, or police captors of the high-profile criminal actors who form the main subjects of these texts. They offer little to the serious academic researcher other than quick, diverting crib sheets or exemplars of the widespread and enduring cultural fascination with the criminal underworld. As with the fascination with the Krays noted below, at times they offer sensationalist or romanticized representations of the underworld sharply at odds with its more prosaic empirical realities (Hobbs and Antonopoulos, 2014: 6).

TRUE CRIME AND GANGSTER AUTOBIOGRAPHIES

A number of colorful, rehabilitated characters formerly active in various underworlds have made decent postcrime careers by retelling anecdotes from their deviant heydays through books, stage shows, and films (Fraser and Morton, 1995; Marks, 1997). These narratives are highly uneven and display tendencies for rampant self-promotion, exaggeration, and romanticism, but do, at times, convey valuable empirical insights into the structures and specific activities of criminal organizations and the operation of criminal markets that might be less readily available from alternative sources (Morselli, 2001). Such accounts, though, should be treated with some caution as routes for researchers wishing to access bygone or enduring underworlds. The pictures they paint are as reflective, often more so, of the agency of these actors in negotiating various extracriminal cultural circuits as they are faithful renditions of specific criminal worlds. The cultural fascination directed toward and the mythologization of London's Kray twins, for example, since they were arrested and imprisoned in 1969, far exceeds their empirical significance within London's overlapping, postwar, criminal economies (Jenks and Lorentzen, 1997).

> The Krays, however, never really made the grade outside of the old East End; as criminal businessmen they were starstruck disasters who were comfortable making their money the old-fashioned way—extortion and intimidation. . . . Yet it was the Krays who attracted the attention of society photographers, and who had contacts in Parliament, the House of Lords and show business, and it is the Krays, in their mimicry of a Hollywood-cloaked mafia, who contrived Britain's most enduring imagery of organized crime: a pretend firm, a make-believe mafia. (Hobbs, 2013: 83)

However, this is not to say that popular or journalistic sources have not been used critically to inform the empirical study of organized crime (Firestone, 1993; Hobbs and Antonopoulos, 2014; Morselli, 2001). For example, Firestone used the growing trend for former mobsters publishing memoirs as sources to explore the motivations of these criminal actors, the organizational aspects of criminal groups revealed through these memoirs, and the growing erosion of trust within American Mafia organizations as evidenced by the publication of Mafia secrets within these volumes. Furthermore, Hall (2016) drew largely on published journalistic accounts from the international financial press in his analysis of the extensive fraud perpetrated by Bernard Madoff. Madoff had been an investment advisor and nonexecutive chairman of the NASDAQ stock market who was arrested in 2008 and convicted the following year of 11 federal felonies related to a massive Ponzi scheme, for which he received a prison sentence of 150 years. The scheme, which relied on paying out existing investors with money derived from later investors in the scheme, had been running since the 1980s and has been estimated at yielding up to US$65 billion (Chung, Daneshkhu, and Masters, 2009). Hall reinterpreted the fraud through a geographical lens, highlighting the social relations of place, in this case New York City and its suburban Jewish communities and social spaces, and the roles that these spatialities played in allowing Madoff to cultivate trust and reputation among the victims he defrauded that was crucial to the enduring success of his criminal activities. Hall added a territorialized interpretation to the wealth of empirical details that were available through journalist sources.

In sum, these popular sources are of some, albeit limited, value to the academic researcher. Many lack the rigor and empirical originality of those investigative journalism sources discussed in the previous section to offer anything to critical students of organized crime. Of value undoubtedly are those studies undertaken by committed investigative journalists such as Anabel Hernàndez, Misha Glenny, Roberto Saviano, and others such as Ed Vulliamy and the host of less-well-known campaigning journalists chronicling unfolding and sometimes violent criminal economies in places such as northern Mexico who are prepared to enter the field, at times at great personal risk and cost, and produce original, primary insights that are not available elsewhere or which would not be accessible to the majority of academic researchers. However, even here the limitations of these contributions should be acknowledged. They are acts of reportage rather than sustained critical historical and theoretical engagements with their subject. It is rare for these contextual elements to be anything more than peripheral aspects of their narratives. And indeed why should they be anything other than this given the more immediate priorities facing the reporter? Rather, building selectively and critically on the more rigorous and original of these works, it is more for the academic researcher to advance these perspectives as part of a complementary endeavor.

SOCIAL MEDIA AND ONLINE SOURCES

A series of emergent sources that cast a variety of contrasting perspectives on regions affected by violent criminal economies, with which there has been relatively little serious academic engagement to date, certainly with specific reference to organized criminal economies and their violent impacts, are those produced by various actors through social media. Most notoriously, social media sites such as YouTube have been used by criminal groups in the exercise of symbolic violence and the generation of propaganda through the posting of videos of violent acts and their aftermaths, such as the killing and mutilation of victims. This has been an observed characteristic of drug-related conflict throughout northern Mexico (Eiss, 2014).

> Dead bodies with signs of gruesome torture, accompanied by *narcomantas* (messages written by the killers on banners) have become ever more frequent. The public display of violence and terror seemingly has

> no limits. Videos posted on YouTube of the torture and killings of employees of rival cartels reveal a sickening and unrestricted climate of impunity and social breakdown. (Watt and Zepeda, 2012: 193)

These incidents mirror the appropriation of social media by groups linked to terrorist acts, which has received somewhat more sustained academic attention (Farwell, 2014; Klausen, 2015; Richards, 2016). Social media have also been employed for the dissemination of other cultural forms supportive of specific criminal organizations or gangsters, such as the posting of *"narco corridos"* (narco ballads) and "narco hip-hop" videos on social media channels. These are musical movements that have arisen in various parts of northern Mexico to praise and glamorize the lives and activities of criminal organizations engaged in drug-related violence there (Edberg, 2001). Equally, within these contexts, there has been a wealth of counter-propaganda narratives produced by citizen bloggers and networks of campaigners and journalists, who have become known collectively as cybervigilantes, who have been active in providing firsthand and reflective testimony of these events (Bunker, 2011a) to counter the propaganda and intimidation originating from cartels.

These online sources offer a potentially rich archive for students of contemporary geographies of organized crime-related conflict. While researchers interested in the study of organized crime have been relatively slow to engage with these sources, they could profitably seek direction from scholars of terrorism who have developed some innovative perspectives with regard to social media sources. These sources have the potential to yield empirical insights into the operation of internal networks of communication within criminal groups. For example, Klausen (2015) employed social network analysis to examine the Twitter accounts of 59 Western-origin jihadist fighters known to be active in Syria and Iraq in early 2014. Klausen detected, that rather than representing a series of unmediated individual narratives, theses Twitter messages actually reflected the controlling influence of feeder accounts from both terrorist organizations within the insurgency zone and of organizations and individuals based in the United Kingdom. Furthermore, social media sources offer resources for the conceptual interpretation of criminal groups as agents within contemporary circuits of communication, technology, and culture. Again, drawing inspiration from critical studies of terrorism, scholars of organized crime could look to Richards's (2016) analysis of the social media activities of the neojihadist group Islamic

State in their acquisition of political-economic capital. For example, there has been little comparable analysis to date of the wealth of social media narratives emanating from the criminal organizations engaged in drug-related conflict in Mexico in recent years. Furthermore, the social media narratives of criminal groups provide evidence for those interested in the more general normative influence of criminal organizations across both specific regional and cybercultural terrains (Midgley et al., 2014: 27; Slade, 2007).

Finally, other online sources, such as the Wikileaks archive, a massive repository of leaked confidential documents, are beginning to the trawled for the potential insights they offer, particularly to our understandings of covert state maneuvering around and discursive framing of the problems of organized crime (Boyce, Banister, and Slack, 2015). All these sources, though, should be approached with a wariness of the issues of state and corporate censorship of these often highly sensitive materials. For example, few cartel killings videos remain on YouTube, in their entirety at least, for very long, for understandable reasons. These, and other online sources, represent fluid, potentially unstable archives, but ones of as yet only lightly tapped value to critical scholars.

HISTORICAL AND ETHNOGRAPHIC ACCOUNTS OF ORGANIZED CRIME

There are a number of further approaches and sources available to critical scholars of organized crime in addition to the ones outlined in the preceding sections. While there are a wealth of standard guides to research methods in the social sciences (Bryman, 2012; Frankfort-Nachmias, Nachmias, and DeWaard, 2014), human geography (Clifford, French, and Valentine, 2010; Hoggart, Davies, and Lees, 2002), and to a lesser extent economic geography (Karlsson, Andersson, and Norman, 2015), there is only a very limited literature specifically designed to inform academic researchers of organized crime (Hobbs and Antonopoulos, 2014). Table 3.1 is based on Hobbs and Antonopoulos's chapter and summarizes the key points from the approaches not previously covered in this discussion.

Critical scholars of organized crime, then, have a broad pallet of methods available to them. These have all contributed in different ways to shaping the recent and contemporary literatures of organized crime. Interventions, for example, from historical and archival analysis have helped add historical validity to critical discourses that have

TABLE 3.1. Additional Approaches to Researching Organized Crime

Aims	Contributions	Limitations	Selected key works
Method: Historical/archival analysis			
Highlighting crime as a historical component of development in the West. Understanding crime within its socioeconomic context. Showing crime to be a malleable concept.	Exploring impacts of historical changes in the law on organized crime. Challenging assumptions about contemporary crime through revealing historical comparisons. Highlighting the multiple links between organized crime and other phenomenon. Highlighting specific human agency.	Not designed to reconstruct organizational and cultural aspects of organized crime. Archives do not provide neutral sources but are reflective of the institutional orders of the time. Occasional tendency of historical sources to romanticize or glamorize their subjects. Inherent issues of authenticity, credibility, representativeness, and meaning.	Bell (1953); Block and Chambliss (1981); Browning and Gerassi (1980); Critchley (2009); Egmond (1993); Haller (1971); Hobsbawm (1972); McMullan (1984); Sharpe (1999); Sherry (1986); Stedman-Jones (1971); Woodiwiss (1988, 2001)
Method: Network analysis			
Sophisticated descriptive approach reconstructing criminal networks through mapping of contacts. Use of various forms of data including wire taps, big data harvesting, autobiographical and biographical sources.	Increasingly sophisticated views of criminal networks, including network strength, differential development of networks, links to legitimate spheres. Challenges myths of hierarchical structural coherence and stability, typical of earlier political and media renditions of criminal organization.	Applications to organized crime still in their infancy. Impossibility of comprehensively capturing totality of criminal connections. Can impose false order or suggest greater network stability than exists within fluid, volatile illicit markets. Picture of underworld of exclusively criminal connections drained of wider cultural contexts.	Campana (2011); Lupsha, (1983); Malm, Kinney, and Pollard, (2008); Morselli, (2005, 2009); Natarajan (2000, 2006)

(continued)

TABLE 3.1. (continued)

Aims	Contributions	Limitations	Selected key works
		Method: Ethnography	
Accessing hidden organized crime populations through extended, intensive immersion in the field.	Revealing the cultural realities of organized crime activities and actors. Greater depth of insight into subjects and their contexts than less directly engaged methods.	Demanding of the researcher in terms of time in the field, emotional investment, social position, capital, and skills and cultural adaptability. Potential dangers to self and ethical or legal ramifications of immersive fieldwork within illicit, sometimes violent, contexts. Validity of ethnographic data and issues of representativeness, replication, and generalizability (common to all organized crime research).	Adler (1985); Bourgois (1995); Brooks (2012, 2015); Cauduro, Di Nicola, Fonio, Nuvoloni, and Ruspini, (2009); Chambliss, (1978); Dunlap, Johnson, and Manwar (1994); Hobbs (1988, 1995, 2013); Ianni and Ianni (1972); Nordstrom (2007); Williams (1989, 1992)
		Method: Interviewing	
Gaining firsthand and reflective testimony from a range of respondents including active, incarcerated, and former criminals; law enforcement agents; victims of organized crime; and members of the public.	Breadth of insider detail gained from interviews with active and former criminals. Highlights patterns of willing consumption of illicit goods and services by the public as well as the predation of victims by criminals. Challenge stereotypical images of victimhood. Comparative perspectives from interviews with different groups/populations.	Differential access to active criminal respondents participating in different illicit and illegal markets. Some market groups very hard to access. Validity of interview data, issues of generalizability and representativeness of sample (common to all organized crime research).	Allum (2006); Antonopoulos, Hobbs, and Hornsby (2011); Decker and Townsend Chapman (2008); Dorn, Murji, and South (1992); Fisher, Ialomiteanu, Russell, Rehm, and Mann (2016); Reuter and Haaga (1989); Ruggiero and Khan (2006); Siegel, van de Bunt, and Zaitch (2003); Wiltshire, Bancroft, Amos, and Parry (2001); Zhang and Chin (2004)

Note. Adapted from Hobbs and Antonopoulos (2014).

located organized crime firmly within, rather than beyond, the contemporary economic and political mainstream. These interventions have revealed historical continuities and highlighted the roles that organized crime has played within various processes of development and state formation. For example, Bhattacharyya (2005: 119; see also Dorling and Lee, 2016: 45–48) has discussed the significance of surpluses from illegal drugs trades in the emergence and development of capitalism historically, while Woodiwiss (2001) has explored the roles of organized crime in the formation of modern America, and Hobbs and Antonopoulos (2013) have explored the normative reimagination and erasure of historical processes of systematic corporate and state banditry and the contributions of these to 20th-century America's emergent institutional order. These and other contributions show that concerns about remapping the place of organized crime within wider economic and political terrains are far from exclusively 21st-century preoccupations.

Ethnographic accounts have produced some of the most telling contributions to recent discussion of organized crime, advancing grounded, peopled, and contextualized knowledges of the operation of criminal economies in various contexts (Brooks, 2012; Hobbs, 1988, 1998, 2013; Massaro, 2015; Nordstrom, 2007). Massaro's account of the "micro-practices of territory, politics and governance" (2015: 369) is a particularly noteworthy example. Combining ethnographic, oral history, and discursive approaches and deploying feminist and urban geopolitical framings, this study examined the discursive construction of one inner-city block in Philadelphia through narratives that render it as a front line in the "war on drugs." It goes on to examine the contestation of these narratives through the lived practices of survival of the block's residents.

However, ethnographic approaches also pose the greatest challenges to researchers seeking to negotiate these fields and their institutional contexts (Brooks, 2014; Garrett, 2013, 2014). Ethnographic approaches speak naturally and inevitably primarily to the local scale. As Chapter 5 will show, this is a scale that has gained relatively little traction in politicized debates about organized crime where multiple perspectives have tended to be submerged beneath or subsumed into discourses of global crime as a series of threats to national territory (Hobbs, 2013: 149–150). Hobbs (1988, 1998, 2001, 2013; Hobbs and Dunnighan, 1998) has deployed intensive ethnographic approaches, reminiscent of earlier phases of sociological enquiry into illicit urban economies (Adler, 1985; Chambliss, 1978; Levi, 2008;

Pryce, 1979; Stedman-Jones, 1971) in his studies of local criminal markets in the marginalized urban spaces of the global North. These have included observation of and interactions and interviews with criminal actors active within the local economies of organized crime. This work aimed to excavate the complex mosaics of local differentiation within the criminal economies of postindustrial urban Britain. The social, economic, cultural, and political characteristics of these localities are incorporated into this work as key variables. Hobbs's work identified characteristics including the nature and organization of the formal economy, neighborhood social makeup, and dynamics and local housing policy as crucial in determining the nature of organized crime in different localities. It was not until these recent empirical investigations, though, that the local dimension of organized crime more fully emerged within its research literatures.

This work has been heralded as theoretically and methodologically significant and can be situated within a wider wariness of overly abstract accounts of globalization. Aas, for example, has argued that "the ethnographic study of culture and cultural variation therefore gains a particular importance as an antidote to the abstract nature of many theoretical claims about globalisation" (2007: 174–175). However, disappointingly, despite the recognition of their significance (Brooks, 2014), such locally grounded ethnographic accounts remain relatively rare. However, perhaps this is not surprising given the difficulties of accessing these economies from an institutionalized position. There is insufficient work in this vein, for example, for us to yet attempt to incorporate such analysis into comparative accounts of the organization, nature, and operation of criminal groups and markets in different places. Other innovative examples of recent ethnographies of the illicit include Nordstrom (2007), who has demonstrated how the ethnographic method can be deployed in a more global sense, and Brooks (2012, 2015), who has employed it within his scrutiny of the illicit in transnational networks of secondhand commodities. This echoes and speaks, to an extent, to the potential applicability of insights from actor–network theory, "follow-the-thing" ethnographies, political-geographies-of-the-object approaches, and object biographies developed within analysis of commodity geographies (Cook, 2004; Cook and Harrison, 2007; Meehan, Shaw, and Marston, 2013) to illicit networks and their transnational materialities. Here methods such as multisite ethnographies have been employed to explore, for example, the connections between sites of production, transit, exchange, and consumption by following the journeys of commodities along the supply chains of a number of licit goods including food (Cook, 2004), money (Christophers, 2011), and fashion (Brooks, 2015), among others. Ian Cook (2004: 642), one of the key pioneers of follow-the-thing methods, argues:

> The research on which it [the follow-the-things approach] is based was initially energized by David Harvey's (1990: 422) call for radical geographers to "get behind the veil, the fetishism of the market," to make powerful, important, disturbing connections between Western consumers and the distant strangers whose contributions to their lives were invisible, unnoticed, and largely unappreciated.

While the applicability of these alternative ethnographic approaches to the commodities of illicit markets is clear, so too are the challenges of operationalizing them across such risky terrains. Despite the many challenges, however, undoubtedly ethnographic approaches offer much potential for the generation of new, nuanced insights into illicit actors, groups, markets, and activities and their impacts across international space.

CRITICAL GEOPOLITICS

A final approach that has been employed to explore the social construction of the illicit through official and popular discourses are a series of approaches grouped broadly under the banner of "critical geopolitics" (Dittmer, 2010; Kelly, 2006; Tuathail, 1996). The starting point of analyses in this vein is the view that discourses, of all kinds, that speak of, in this case, the illicit, are rarely neutral and value-free, even when, and perhaps especially so, when they claim to be so. Rather, these discourses are inscribed with values and ideologies that reflect the institutional positions of their authors. The task of critical geopolitics is to deconstruct the various texts that constitute these discourses and to reveal these, at first, hidden intentions. Scholars working within critical geopolitical traditions are also interested in the power of texts in terms of their use and consumption. Thus, the deconstruction of texts is often accompanied by consideration of the circulation and consumption of these texts and their effects among a variety of audiences. This approach offers the researcher the advantages of a field of little risk that is populated largely by readily accessible sources. Anecdotally, though, it would appear that there is a bias toward Anglophone sources within work in this vein. The limitations of the field largely revolve around the dangers of advancing naïve or self-indulgent readings of sources that connect little with the political or lived realities of illicit activities.

There is a wealth of sophisticated scholarship within this tradition that has scrutinized a number of popular and official texts that reveal a variety of value-laden geopolitical visions that are often hidden beneath the veneer of popular entertainment or official neutrality (Dittmer, 2007, 2010; Dodds, 2005; Moore and Perdue, 2014; Pickles, 2004; Wood, Fels, and Krygier, 2010). There is also an emergent body that is interested directly in texts that speak of the illicit and activities associated with organized criminal markets. These include Jones's (2014) reading of the American television series *Border Wars*. The popular appeal of the program hinges on the imagination of threat from beyond the United States–Mexican border around narratives of encroaching conflict, illegal migration, and drug trafficking. The series undertakes this imaginative labor despite the paucity of on-screen empirical evidence to sustain its thesis and a lack of contextualization of its subjects within the wider geopolitics of the region. Furthermore, Dodds (2013) explores the geo- and biopolitics of illegal migration and the materiality of borders at the United States's northern border with Canada in his reading of the 2008 film *Frozen River*. There is a wealth of other such texts that represent aspects of the illicit generally and organized crime specifically that are ripe for such critical examination. While scholars of critical geopolitics have readily engaged with popular texts, similar critical engagement with official texts and the discourses they constitute would advance scholarship considerably in this area. Chapter 6 provides a brief example of this possibility in its exploration of official discourses of the multiple criminal mobilities that have become framed through a politicized metaphor of "flow."

POLICYMAKING AND KNOWLEDGES OF ORGANIZED CRIME

Academic knowledges have made relatively little impact in shaping policy responses to organized crime. The increasingly nuanced understandings of organized crime that they offer have not necessarily resulted in more nuanced policy responses to the problems it poses. This reflects both a reluctance of policymakers to seriously engage with academic findings and a failure of many academic discourses to connect with the priorities of the policymaking community, address gaps in their knowledge, or indeed to address them in appropriate idioms (Jesperson, 2014: 154). It has been said that academic knowledge enjoys only a "lowly status" (Hobbs and Antonopoulos, 2014: 2) and that academic researchers and policymakers have proceeded on "different tracks" (Jesperson, 2014: 154). Policymaking around organized crime has long been criticized as reacting more to conspiracies and moral panics, typically constructed around one or more supposed alien threat (Hobbs, 1998, 2001, 2013; Madsen, 2009; Woodiwiss and Hobbs, 2009) than to the findings of independent research (Hobbs and Antonopoulos, 2014: 1–2). This may help explain why the majority of policies designed to eliminate, reduce, or manage various activities' links to organized crime have been such spectacular failures (see Chapter 7).

Academic explorations of organized criminal markets should not be restricted to the pursuit of instrumental knowledges to serve the policy community. It is from a critical, independent position that many of its most insightful contributions have emerged, many of which do not have any immediately obvious policy application. However, it has been argued that a more extensive dialogue between the policy and academic communities would be both timely and profitable (Jesperson, 2014: 154). To this end in the United Kingdom the independent think tank the Royal United Services Institute launched the Strategic Hub on Organized Crime in association with the U.K. government's Home Office, National Crime Agency, Foreign and Commonwealth Office, and the U.K. Research Councils' Partnership for Crime, Conflict and Security. The Strategic Hub on Organised Crime is described as "platform for policymakers to articulate their needs for research and analysis and for academics to share their areas of work" (Jesperson, 2014: 154–155). While such dialogues around issues of organized crime have been limited historically, there have, in recent years, been growing imperatives inscribed within funding regimes for academic research to increasingly demonstrate its impacts on society. There is some evidence that academics are intervening more noticeably within such policy debates. For example, there are cases of some national and state governments in places such as Portugal in Europe and the states of Alaska and Colorado in the United States retreating from the strident prohibition of some narcotics that had previously prevailed internationally. There has been a wealth of academic interventions around these developments and wider debates about drug policy that have either sought to evaluate the impacts of changes in drug policy (Hughes and Stevens, 2010; Uchtenhagen, 2009), assess public opinion (Fisher, Ialomiteanu, Russell, Rehm, and Mann, 2016), or advocate for more widespread policy revisions (Collins, 2014b). While it is not yet clear how influential these academic voices have been within an apparently changing policy landscape, it would appear that the priorities of the academic and policy communities, in these instances at least, are more clearly aligned. Challenges for researchers in this area in the future will revolve around their negotiation of these imperatives to produce both applicable and independent critical knowledges.

Summary

This chapter has highlighted the reality that researching organized crime, whatever the purpose and approach of this research, is a far from an easy, straightforward, and unproblematic endeavor. While there is a very wide pallet of perspectives from which to approach the field that are available to the researcher, all throw up their own very different practical, ethical, and definitional challenges. There is a now a long-standing official tradition of attempts to measure and estimate, in various ways and within various contexts, the extent of organized criminal activity. Such publicly available sources of data, notwithstanding their obvious flaws and limitations and the politicized ends toward which they have often been deployed, offer resources for the shaping of alternative critical narratives of organized crime. Similarly, journalistic, biographical, true crime, and social media sources offer either insightful, all be they somewhat uneven, complements to other knowledges of organized crime, or, alternatively, important sources for critical scholars to engage with.

Strictly academic knowledges of organized crime have had, and continue to have, much to say, providing a diversity of new knowledges that frequently challenge many taken-for-granted assumptions and stereotypes of organized crime. While vital, it is disappointing to see what little traction these knowledges have enjoyed within policy and law enforcement circles. The fault here lies on both sides. Academics, for example, in some cases continue to view law enforcement agencies and government policymakers and the discourses that they produce with some suspicion or as the specific objects of their critical scrutiny. Despite the long-standing bifurcation of these academic and applied discourses, there appears to be a move within contemporary discussions of organized crime advocating a more mutually profitable dialogue. This is to be welcomed, especially if it produces more informed policy stances.

Beyond the development of more sustained academic–policy dialogues around the questions raised by organized crime, this chapter has highlighted a number of ways in which attempts to measure and research organized crime might be advanced. First, the definitional confusion highlighted at the start of this chapter should be addressed. Work might profitably be undertaken at an international level to develop a standardized set of definitions for organized crime and its various markets. These should then seek to inform the collection of data by official agencies in different countries. While it is unlikely that such definitions will be perfect or necessarily universally agreed upon, such an endeavor should work toward evening out some of the problems caused by different definitional and collection practices that currently bedevil the study of organized crime through official statistics. At least there will be more common understanding of the phenomena under investigation. It should be acknowledged, however, that addressing differences in the entrenched national reporting cultures of official agencies will be more difficult and they will take longer to shift toward more standardized, international models. These standardized definitions, then, should refer to all aspects of the economies of organized crime. Unfortunately, experiences of attempts to standardize international practice in the areas of anti-organized crime initiatives does not bode well for such a project. At least working toward definitional standardization should prove less contentious and damaging than some attempts at regulatory or procedural standardization (Sharman, 2011).

Two areas of research in the collection of organized crime data where critical scholars might pioneer are standardized methods for collecting data on the overlaps and connections between criminal organizations and economies and licit corporations and markets, and the question of gathering data on the potentially (and arguably) developmentally positive impacts of illicit markets of organized crime. Given the traditionally politicized renditions of organized crime that have emerged from official sources, it is unlikely that official agencies would be willing to pioneer such nuanced and potentially disruptive data lenses.

Finally, it is worth reemphasizing the potential of ethnographic approaches to advance understandings of organized crime, not just at the local scale, at which they have tended to be applied in the past, but in a comparative and transnational sense through the application of creative approaches such as the multisite ethnographies typical now of follow-the-things approaches. This chapter, then, has demonstrated that there is considerable potential for us to enhance both our quantitative and qualitative knowledges of organized crime and that this should be an immediate methodological research priority for scholars and students of organized crime from whatever background.

CHAPTER 4

THE ORGANIZATION OF CRIMINAL ENTERPRISES

There has been a change from gang-related "underworlds" of the pre- and post-Second World War periods to a more fragmented and diverse panorama of criminal groups at the beginning of the twenty-first century.
—WRIGHT (2006: XVII)

INTRODUCTION

Economic geographers have long been interested in many aspects of the organization of economies. These have included, for example, the structures of enterprises and corporations, their endurance and change over time, and their regional and international variations; the effects of external (regional, national, and international) contexts on the organization of economies and the enterprises that populate them; the multiple forms of regulation that exist in markets such as legally binding contracts, trade associations, and widely agreed standards, as well as the less formal roles of trust and acquaintance in some markets and how these affect the nature, operation, and spatialities of markets; and the relationships between enterprises, between enterprises and labor, consumers, institutions, and the state. Some of the key insights to come out of this work have included the observation of the emergence of polarized organizational forms in contemporary markets reflective of the regime of flexible accumulation, stability, change, and diversity in forms of economic organization; the recognition of national business cultures and their associated forms of business and corporate structure and behavior; regulatory forms and asymmetries and their effects; and organizational structures that reflect specialisms and divisions in the labor process and their expression across uneven global space (Coe, Kekky, and Yeung, 2007; Hudson, 2005). This chapter explores a number of aspects of the organization of criminal economies, asking similar questions to those pursued by economic geographers in their analyses of licit markets.

"Criminal organization" here is taken to include characteristics such as the size, structure, basis for membership, regulatory resources, and intergroup relations of criminal groups. This chapter situates criminal enterprises in their wider economic contexts, tracing the interdependencies between licit and illegal markets. In doing so, it will critically review a number of claims commonly made about the organization of criminal enterprises, which have emphasized qualities such as the rationality, hierarchical organization, size, tightly bound intergroup relations, transnationality, mobility, and monolithic natures of criminal organizations. It also explore examples of the basis for membership in criminal organizations, focusing particularly on the enduring significance of place; the regulatory resources deployed by criminal groups in attempts to maintain order and ensure exchange in the economies within which they operate; and the evolution of criminal organization within the context of post-Fordist capitalism. Throughout, the chapter questions the ontological separation of licit, illicit, and outright criminal markets within discussion of questions of economic organization and what this might say about some of the limitations of the broader academic literatures on the organization of economic enterprises.

This chapter aims to contribute to contemporary understandings of the organization of criminal economies in a number of ways. It explicitly focuses on the roles of space as a dimension of the nature of criminal enterprises, their structures, and organization. It considers here, for example, the role of place as a basis for organization of criminal enterprises; differences in the nature of criminal organizations observable across space; and the challenges that space poses for the regulation and mobilization of criminal resources of various kinds, and hence its potential effects on the size, scope, and nature of criminal organizations and the relationships between criminal groups. The chapter explores the notion that space can be both a resource to criminal organizations in that they are able to draw strength from their embeddedness in particular regions and also a challenge in that it has proven far from straightforward for criminal organizations to mobilize beyond their regions of origin.

FORMS OF CRIMINAL ORGANIZATION

Criminal markets are characterized by a diversity of organizational forms. One of the key determinants of criminal group structures are the activities in which groups are engaged. "Different criminal activities call for different structures" (Serious Organised Crime Agency, 2006: 54). Small, tightly knit groups composed of specialists are characteristic of activities such as bank robbery that require coordination, speed, and agility (Serious Organised Crime Agency, 2006: 54). The heist of an estimated £14 million of gold, diamonds, jewelry, and cash from the Hatton Garden Safe Deposit Company in central London in April 2015, for example, was carried out by a gang consisting of seven or eight London-based career criminals, with ages ranging from 48 to 76, all of whom demonstrated long involvement in the city's violent criminal economies. The history of major armed robberies in London had been typically characterized by gangs of a similar composition, although recent crimes have demonstrated a more multicultural character (Campbell, 2016). Other small criminal groups, such as those observed by Stephenson in Russia, who operate in less-specialized economies but who are more generally dedicated to material accumulation, tend to also be tightly bound but with members who display less differentiated functions. Here the associations between members are more likely to be based around a common territory rather than complementary criminal skills and experience (Stephenson, 2015: 105). They are also more enduring than the project-based associations characteristic of heist economies.

Smuggling, in contrast, requires groups who are more cellular and networked in nature, with different cells undertaking different functions (Hignett, 2005; Levi, 2002; Madsen, 2009: 7–8; Serious Organised Crime Agency, 2006: 54; Wright, 2006: 7). Both cellular and networked group structures have also been detected within the economies of cybercrime (Glenny, 2008: 306; Serious Organised Crime Agency, 2006: 54), although there is evidence of some larger, more corporate forms developing in exceptional cases as the scale of cybercrime has developed with time and the opportunities afforded by technology. Glenny (2011), for example, discusses the case of the Ukraine-based enterprise Innovative Marketing, which perfected the "scareware" cyberscam in which computers would become infected through bogus adware distributed by the company. Infected computers then directed their users to buy useless software from Innovative Marketing in the belief that it was the only way to rid their computers of the virus it had acquired from Innovative Marketing's adware. The enterprise had all the appearance of a legitimate corporation, so much so that many young people working for it in Kiev believed they were working for a legitimate enterprise. The company also ran three international call centers to assure customers who were trying to install the useless software of the company's legitimacy. In total, it is estimated that this scam generated tens of millions of dollars for the managers of Innovative Marketing (Glenny, 2011, 173–174).

An organizational form, dominant within the licit economy, which has little equivalence within criminal economies, despite cultural discourses that often stress the contrary, is the large, multi-, or transnational corporation. To a large extent the contemporary licit global economy is an economy of such corporations, who account for over 70% of world trade in the licit sector (Steger, 2003: 48–49). However, there are very few examples of criminal organizations that could be classed as large and none that have the global reach typical of many licit transnational corporations. Extensive cross-border trafficking economies can, in some instances, become characterized by large criminal organizations capable of controlling extensive areas of territory and deploying their own armed militias (Saviano, 2016; Serious Organised Crime Agency, 2006: 54). Despite their centrality to many political, security, and cultural discourses of organized crime, though, such organizations remain rare and require exceptional circumstances within which to develop (Watt and Zepeda, 2012).

Extensive criminal organizations are found in both southern Italy (Levi, 2014: 11) and Mexico (Watt and Zepeda, 2010). In the case of Italy, the antecedents of present mafia associations can be traced back to at least 1880 where, at times, they have enjoyed particularly close associations with the Italian state. The complexity, longevity, and degree of political embeddedness of these criminal organizations have been glimpsed in only a very few other cases (Paoli, 2005: 20–21). In the case of Mexico, the emergence of major drug-trafficking cartels is a much more recent phenomenon. Here such groups have emerged out of long-standing and deeply rooted corruption within the Mexican state that has spawned enduring relations of complicity between organized crime, politics, and business there. This cartelization is also, to an extent, an unforeseen outcome of the

U.S. supported antinarcotics initiative, Operation Condor (Watt and Zepeda, 2012). In eliminating a number of small trafficking organizations, Operation Condor reduced competition within the northern Mexican trafficking "plazas," which allowed the development of a small number of very large narcotrafficking organizations. Midgley and colleagues argue, "Market interventions by states disturbs the political economy of the trade, cultivating more violent actors, in turn driving more aggressive state interventions which in turn drive more violent outcomes" (2014: 10). This concentration was aided by corruption that selectively favored the large, well-resourced trafficking organizations (Watt and Zepeda, 2012: 51). A similar process of cartelization, as an unintended consequence of interdiction efforts, has been observed in Afghanistan: "In Afghanistan, interdiction efforts of the mid-2000s that focused on the least powerful small traders led to a vertical integration of well connected drug capos and enabled the Taliban to reintegrate itself into Afghanistan's drug trade" (Felbab-Brown, 2014: 46; see also Goodhand, 2009).

At the start of the narcotics supply chain that runs through Mexico sits the Latin American, and specifically Colombian, cocaine industry. Columbia has long been recognized as one of the world's leading producers of cocaine (Alexander, 2015; Bagely, 2005). A number of trafficking organizations, most famously the Medellín cartel led by Pablo Escobar prior to his death in 1993, have been associated with the rise of Colombia as a cocaine-producing and exporting state from the 1970s. Early analysis of the Colombian cocaine industry suggested an industry dominated by extensive cartels (Castells, 2000). However, later scholarship has thrown doubt on both the size and the structural coherence previously attributed to these organizations, identifying them, rather, as more akin to fluid, flexible networks (Kenney, 2007: 233). Recent iterations of the industry there have been identified as organizing along smaller "*cartelito*," or microcartel, lines (Bagley, 2005: 39; Kenney, 2007: 259; Saviano, 2016: 159). Even in Mexico there is some evidence of structural disintegration within the cocaine trafficking economy and the emergence of some microcartels:

> The repression carried out in recent years by Mexican law enforcement and American intelligence against groups that, until a short time ago, were major cartels has often led, not to the annihilation of these organizations, but to the metastatic proliferation of microcartels. These are smaller structures that are born out of struggling mother organizations through the exploitation of power vacuums. (Saviano, 2016: 105)

It is common for criminal organizations to be active in multiple criminal markets, especially where they are long established or large in size. Europol (2013: 33), for example, argues that "more than 30% of the [organized crime] groups active in the EU are poly-crime groups, involved in more than one crime area. Almost half of these poly-crime groups are linked to drug trafficking, and 20% of these groups engage in poly-drug trafficking." The literature records many examples of criminal groups such as Italy's mafia associations, Japan's Yakuza, or Jamaica's Shower Posse's involvement across a range of licit and illicit markets including the trafficking of numerous illicit commodities, the control of prostitution, money laundering (although this is a necessary adjunct to almost any successful criminal activity that generates significant illicit funds), protection and extortion, control of illegal gambling, property development, corruption of the provision of environmental management services, counterfeiting, and corruption (Glenny, 2008; Hill, 2005; Kilcullen, 2013; Madson, 2009; Paoli, 2005; Saviano, 2008; Serious Organised Crime Agency, 2006). As Kilcullen (2013: 92) argued of Jamaica's Shower Posse, a powerful, transnational Kingston-based criminal group, led by Christopher "Dudus" Coke until his arrest and extradition to the United States in 2010, "drug trafficking doesn't define what an organisation like Coke's group is; it's just one of the things the group does."

Criminal organization cannot be simply reduced to being the product of economic rationality. To do so carries the risk of reductionism, overlooking the social dimensions of criminal group organization. Despite frequently being referred to as such within media and some academic discourse (Williams, 2001: 106), criminal groups are not merely businesses who differ from corporations only in that they are largely active in illicit rather than in licit markets. Despite this reality, the literatures of organized crime have tended, traditionally, to offer somewhat restricted takes on the logics of group organization, emphasizing either their commitment to material accumulation (Carter, 1997; Dick, 2009; Ruggerio, 2009; Smith, 1980), the social bonds of their members (Albini, 1971; Ianni and Ianni, 1972; Lombardo, 1997), or their control of specific territories. More recent empirically grounded accounts of criminal organizations, drawing on a range of data types derived from ethnographies, interviews, and analysis of criminal case files, have emphasized the idea that criminal organizations are both economic and social forms (Hobbs, 2013; Kilcullen, 2013; Kleemans and van de Bunt, 1999; Stephenson, 2015).

Criminal organizations, then, routinely combine internal moral commitment with material accumulation. Stephenson (2015: 94) noted of the groups she observed that "the Russian bandit gang is neither a purely social organization nor an illegal business corporation but a multifunctional clan." This multiple nature of criminal organizations is discernible within the internal structures of many criminal groups. Rather than the arboreal, tree-like, hierarchical structures than many have attributed to criminal associations, they typically display a flatter, fluid, more egalitarian, dispersed, rhizomatic structure. Here authority within these organizations is likened not to that of the corporate business manager, but to the ritual specialist who "makes sure that the gang remains a cohesive moral unit" (Stephenson, 2015: 105; see also Hallsworth, 2013).

The preceding discussion of criminal organization carries with it the considerable risk of overgeneralization, given the enormous social and geographical territory across which it speaks. It should not be forgotten that there is great variation in the organization of criminal groups. It is to the detail of some of these variations and to recent development in the organization of criminal groups that the chapter now turns.

CRIMINAL ORGANIZATION UNDER POST-FORDISM

It is clear that organised crime is going through a period of rapid and dramatic change. Globalisation is reshaping the underworld, just as a combination of evolving law-enforcement strategies and technological and social change is breaking down the old forms of organised crime (monolithic and identified by physical "turf" or ethnic identity), and creating new, flexible networks of criminal entrepreneurs.
—GALEOTTI (2005A: 5)

It is becoming increasingly axiomatic across a diverse and multidisciplinary set of literatures for authors to assert that organized crime is not restricted to a discrete economic realm, but that it overlaps and is interconnected with a number of licit economic realms (Hobbs, 1998, 2001; Hobbs and Dunnighan, 1998; Hornsby and Hobbs, 2007; Lavezzi, 2008; Ruggerio, 1995, 2009; Scalia, 2010; Weinstein, 2008). Thus, we should not think of criminal organization as a hermetically sealed, unchanging, discrete realm unaffected by changes occurring within the economic mainstream. Rather, it should be thought of as dynamic and reflective of its economic contexts, relationships that are observable across a variety of scales (Hudson, 2014: 780). At the macroscale a number of authors have pointed to temporal dynamics in the nature of capitalism, namely, the emergence of neoliberal, post-Fordist, globalized regimes of accumulation as affecting a number of changes in the organization and nature of criminal markets and the groups associated with them (Foltz et al., 2008; Hobbs, 2001; Scalia, 2010; Stephenson, 2015).

There is an extensive literature that has scrutinized the transformations of global capitalism since the 1970s into a regime of flexible accumulation (see, e,g., Amin, 1994; Dicken, 2011; Harvey, 1989; Kiely, 1998; Lash and Urry, 1987, 1993; Monbiot, 2016). Despite the diversity of disciplinary and theoretical perspectives and empirical contexts characteristic of this literature, there is broad agreement that economic conditions across the variegated spaces of the global economy (Peck and Theodore, 2007) have generally become less certain and predictable; more volatile in terms of growth, both through time and across space; and more intensely competitive globally (Harvey, 1989). This, in turn, has affected responses in the organization of firms and in their production practices, which have been observed as becoming less vertically integrated, more flexible, and more responsive to increasingly rapid shifts in consumer demand. Much discussion within this vein from these literatures has explored the organizational advantages offered by extensive transnational corporations and, in contrast, small- and medium-sized enterprises within these contexts. While transnational corporations are able to achieve advantages through a dispersed organizational structure that exploits spaces of flexibility that have opened up internationally, small- and medium-sized enterprises are more suited to exploiting short-term or highly specialized consumer niches or responding with agility to unexpected spikes in demand (Baker, 1993). Commentators have also observed firms articulating with their workforces in increasingly diverse and flexible ways and employing practices such as the use of illegal labor, for example, drawn from the informal "sweatshop" economies of the cities of the global South (Daniels, 2004), project-specific consultants, the increasing use of short-term and zero-hours contracts, and other forms of flexible labor practice such as "hire and fire" or "off the books" work (Harvey, 1989).

There is broad agreement within the criminological literature to suggest that the organization of criminal groups has undergone changes comparable in nature to those observable within the licit economy. Many authors have, in recent years, recorded examples of criminal organizational forms that appear to offer entrepreneurially oriented, opportunistic flexibilities, advantageous within the more disorganized economies, licit and illicit, within which they operate, and which mirror, to some extent at least, characteristics of firms found in the licit economy within this period. These include the apparent preponderance within many criminal markets of small, dynamic, ephemeral, networked, project-oriented, loose collectives of criminal actors (Galeotti, 2005a; Glenny, 2008; Hobbs, 2001, 2004; Hudson, 2014; Kleemans and van de Bunt, 1999; Scalia, 2010; Serious Organised Crime Agency, 2006; Stephenson, 2015; Wright, 2006). Within these broad descriptions, though, there exists considerable variation, some of which is regional in nature, a question that is returned to later in this chapter (van Duyne, 1996; Wright, 2006: 20). As Levi (2014: 11) observes, "aside from Italy, mafia type large scale organised crime is rare in the EU: looser affiliations, akin to those of small and medium sized enterprises (SMEs) are far more widespread." Such smaller criminal affiliations are often contrasted to the more rigidly organized crime groups characteristic of earlier periods where territory, kinship, and ethnicity were central to their organizational logics, their exclusivity, and also their social embeddedness across their territories.

Although there is general agreement that changes are afoot in the organization of criminal markets, it is less certain to what extent these changes represent a fundamental or wholesale transformation to plural, fluid models of criminal organization. It is undoubtedly true, something that is very evident within the criminological literature, that, in many instances, the role of place and ethnicity is less central than they used to be in the organization of criminal groups (Kleemans and van de Bunt, 1999: 25; Nordstrom, 2007: 144; von Lampe, 2012: 192–193). For example, Galeotti (2005b: 58) discusses the growing multiethnicity of Russian crime groups: "As for ethnic ties, even the so-called 'Chechen mafiya' now includes Georgians, Dagestanis, Kazakhs and even Slavs." However, it would be naive to imagine, for example, that criminal groups whose membership was based around the supposed traditional organizational logics of place or ethnicity have vanished. By contrast, the literature continues to record the persistence of many such groups (Grascia, 2004; Hill, 2005; Paoli, 2005; Schoenmakers et al. 2013; Stephenson, 2015; Varese, 2011; Wainwright, 2016: 80–81, 83), while Hobbs (2013: 157) maintains that kinship and ethnicity do remain key means of ensuring trust within criminal economies and Kleemans and van der Bunt (1999) make the point that ethnicity remains important if not now determinate in the shaping of criminal organizations. Evidence from analysis of organized crime networks in the Netherlands suggests that ethnicity remains an important dimension of criminal groups, but that it is less determinate than other factors that bind criminals together (Schoenmakers et al., 2013: 329–330). Of criminal groups in the Netherlands, Kleemans and van de Bunt (1999: 25) argue:

> Ethnicity certainly affects the composition of criminal associations. When offenders, however, cooperate with people of the same ethnicity, it is not ethnicity that matters in the first place, but the fact that these people are family or originate from the same village or the same region. The basis of criminal associations is not formed by ethnicity, but by the social relations that exist between various persons. This applies to both immigrant and native offenders. . . . Hence ethnicity is a significant factor, as it affects social relations. It just so happens that family and friends often have the same ethnicity.

Furthermore, Reuter (2014), drawing on Arlacchi (2004) and Paoli and Reuter (2008), discusses the importance of diaspora populations in shaping transnational drug trafficking networks within Europe, suggesting that ethnicity remains of importance not just to the internal organization of criminal groups, but perhaps also to the brokering and maintenance of relations between different groups within criminal networks. What these discussions of ethnicity should not blind us to, however, is the evidence that ethnicity might be a "superficial characteristic of criminal networks based on family, friendship, or local community ties" (von Lampe, 2012: 193). Von Lampe goes on to cite evidence, albeit somewhat anecdotal in nature, that criminal network ties can be forged in the absence of such strong social and cultural ties and commonalities, but through connections such as childhood acquaintance, legitimate business operations, and, occasionally, through social gatherings of criminal actors (von Lampe, 2012: 193).

The evidence concerning the enduring significance of territory to the definition and organization of criminal groups is somewhat mixed at first reading. A number of authors have argued that as populations have become generally more mobile, the influence of territory on criminal group organization has declined somewhat and, in much the same way as it has been argued that globalization has led to some dilution of territorial cultures and identities, that the regional identities of criminal organizations have similarly weakened as they have also become more mobile (Galeotti, 2005b; Hobbs, 2004; Nordstrom, 2007: 13–14; Serious Organised Crime Agency, 2006). Such viewpoints lend credence to the argument that changing criminal economies mirror, in both cause and outcome, recent changes occurring within the licit economy that are associated with the processes of globalization (Shelley, 2006: 43). However, there is evidence derived from extensive empirical studies that suggests territory retains an importance both to the definition and continuation of criminal organizations. For example, Varese (2011), with reference to a series of case studies of criminal organizations in Italy, Russia, the United States, and China, demonstrates the difficulties they face in migrating beyond their home region and successfully transplanting or expanding into new regions (see also Gambetta, 1993; Reuter, 1985). Perhaps, then, the mobility of criminal groups is somewhat overstated. Varese's work shows these groups to be anything but weightlessly mobile across global space and offers a convincing refutation to claims, discussed in Chapter 1, of the recent emergence of *pax mafiosa* (Jamieson, 1995; Robinson, 2002; Shelley, 2006; Sterling, 1994). Furthermore, Stephenson's research into post-Soviet Russia gangs found them typically to be the products of specific places and she felt able to argue "generally, the gang is a territorial organization" (2015: 129), at the lower levels at least. Her research did, however, reveal that the criminal associations of gang leaders who emerged in Russia in the late 1980s and early 1990s as the Soviet Union collapsed and who accumulated great material wealth moved across economic sectors and wider geographical space, transcending their associations with specific urban turfs (Stephenson, 2015: 92).

This is not to say that organized criminal groups are immobile or unable to migrate and become active internationally. Indeed, Varese (2011: 5) cites the examples of the successful migration of the Calabria-based 'Ndrangheta in the 1950s to Piedmont in the north of Italy and the migration of Italian mafia to the United States in the 19th century. There are also many examples of domestic criminal organization being present and active within their international diaspora communities (Galeotti, 2005b; Lintner, 2005). Rather, this evidence suggests that the mobility of criminal organizations is more fettered and more difficult to achieve than is often supposed and has not occurred to the extent that we can now discount territory in our attempts to understand the organization of criminal groups. Thus, while territory, ethnicity, and kinship might be becoming only some among a growing number of bases upon which criminal groups are defined and organized (Nordstrom, 2007: 144), they have far from disappeared (Siegal et al., 2003). While geographical differences between gangs might not be as pronounced as they have been previously, criminal economies have not yet become, nor are they showing any signs of becoming, homogeneous, placeless, or truly cosmopolitan social terrains.

At the regional and urban levels, a number of demographic changes linked to the mobility of many contemporary societies, in places exacerbated by local housing policy, have conspired to make metropolitan criminal markets more diverse terrains, a mirror of their increasingly plural urban settings (Aas, 2007; Abraham and van Schendel, 2005; Bhattacharyya, 2005; Hobbs and Dunnighan, 1998; Madsen, 2009; Passas, 2001). However, this is neither universal, nor does it necessarily produce more multicultural criminal organizations. Rather than the emergence of more generally multicultural and integrated criminal economies, it would appear that the emergence of new markets, particularly those associated with transnational drug trafficking and consumption, have necessitated greater intergroup cooperation than was typically the case in earlier, more locally oriented, criminal economies (Hobbs, 2013; Levi, 2008; Schoenmakers et al., 2013; Silverstone and Savage, 2010).

Thus, it would appear that criminal groups are able to achieve flexibilities within organizational forms that, at least in part, continue to demonstrate supposedly traditional characteristics. It is important, therefore, to avoid the trap of organizational determinism when discussing the recent dynamics of criminal organization. A case in point would be the Vietnamese crime groups associated with the expansion of domestic cannabis production in the United Kingdom since the drug was reclassified in 2004. While remaining rigidly monoethnic, these groups have developed an organizational form well suited to the market conditions within which they operate. For example, their cannabis economies are characterized by a highly specialized division of labor, the brokering and maintenance of relationships with criminal contacts of other ethnicities, highly mobile investors who may be involved in cannabis operations in many parts of the country, and networked forms that, while displaying some degrees of permanence, are also adaptive to changing circumstances when necessary (Silverstone and Savage, 2010: 29). Schoenmakers and colleagues (2013: 330) observed similar organizational traits in the case of Vietnamese crime groups operating within the cannabis market in the Netherlands, noting, "For specific tasks ... use is also made of co-offenders from a different origin and there is cooperation with other groups to sell the cannabis."

Where, however, for whatever reason, criminal groups, in their organization, remain overly rigid or disengaged from the economic and social mainstream, they tend to struggle to adapt to and thrive within their changed economic circumstances. Such organized criminal groups do endure in the 21st century, but they tend to do so while occupying more marginal positions within contemporary criminal economies (Hobbs, 2001, 2013; Stephenson, 2015). Often the weighty cultural capital still clinging to the reputations of these longstanding groups is sharply at odds with their increasingly peripheral locations within the true political economies of crime. One such example is the *vory v zakone*, the Russian criminal gangs born deep in the prison systems of the Soviet Union. The closed contexts from within which the *vory v zakone* emerged created close allegiances between its members based around rigid, unbending hierarchies and a code that deliberately cut them off from mainstream society. Whereas this was an appropriate response to the circumstances of their creation and development and has been successful in building them considerable and enduring cultural capital across many sectors of Russian society, it left them vulnerable within a market system that values connections with the mainstream and that makes a virtue of flexibility. Stephenson contrasts the responsiveness of the newer street banditry of the post-Soviet era (the "lads") with the now creaking relics that emerged from the Soviet prisons: "In the lads' descriptions, their code is constantly evolving, unlike the immovable, highly prescriptive law of the vory. Fundamentally unchangeable traditional authority is not for them. Instead they are flexible, pragmatic and open to new challenges" (Stephenson, 2015: 185). By contrast, of the *vory v zakone* Stephenson (2015: 184–185) argues:

> Generally, the vory v zakone society, with its historical prohibition on contacts with the outside world (particularly agents of the state), found it difficult to adjust to a new reality. Capitalist accumulation demanded deep penetration into the corporate and state structures, and the new bandits, whose code . . . allowed and even encouraged wide social connections of their members, had a great advantage over the rigid and closed society of the vory.

Furthermore, Hobbs and Dunnighan discuss cases of the decaying family firms they observed in their fieldwork in declining former industrial cities in the United Kingdom. These firms, they argued, were trapped within the declining formerly industrial zones and the increasingly fragmented working-class communities of Britain's postindustrial cities (1998: 291–292). In such circumstances these groups either adapt or become relics:

> The fragmentation of both traditional working-class neighbourhoods and the local labour markets upon which they were dependent . . . made it difficult for [the] family based units . . . to establish the kind of parochial dominance they once enjoyed (Hobbs, 1995: 67; Stelfox, 1996: 18). As a consequence, working-class crime became relocated more than ever within ad hoc, loose-knit, flexible, adaptive networks whose scope exceeded circumscribed terrain and which were characterized by an ability to splinter, dissolve, mutate, self-destruct, or simply decompose in response to the uncertainties of the marketplace. (Hobbs, 2013: 94)

Elsewhere others have observed the social and cultural impacts of economic transformations on criminal groups through the growing tensions between the symbolic and ritualistic expressions of organized criminal groups as collective entities and the encroachment of an increasingly capitalistic ethos into criminal groups who have become characterized by growing internal economic differentiation. For example, Stephenson's (2015) reading of the evolution of the Russian street gang under post-Soviet capitalism records the growing separation between younger members whose economic orientation remains, often through necessity, toward the worlds of the street, and older members who have secured greater reward through their partial assimilation into legal economic realms, albeit often through initially predatory avenues. This, she and others have noted, has initiated the partial decay of these gangs' fraternity under the growing influence of an entrepreneurial group orientation (Paoli, 2005; Wright, 2006: 1).

> But the gang's Achilles heel is the social differentiation that arises in the process of capitalist accumulation. When organized crime moves away from its territorial roots, solidarity comes under serious strain. With the leaders increasingly seen as betraying the ideals of brotherhood and surrounding themselves with paid associates, their form of traditional patrimonial authority loses its legitimacy. As the economic activities of the gangs' leaders progressively acquire the character of rational business, the practices and rituals that sustained collective identities lose their magic. Ordinary gang members can feel exploited and alienated. This is exacerbated by the fact that prospects for serious social mobility through the gang alone are long gone. (Stephenson, 2015: 126)

One way in which the small, flat, ephemeral criminal group form that has emerged under post-Fordist conditions is not obviously suited to the markets they operate within is their ability to organize their operations internationally. All the key trafficking markets that now dominate criminal economies around the world are transnational to some, and typically to a large, extent. Within the licit economy the global articulation of economies is achieved through the extensive transnational form of large corporations. The multiple functions associated with these corporations are dispersed internationally and these corporations organizationally span these spaces and develop transnational supply chains internally (Dicken, 2011). In having no such equivalents within criminal economies, these markets must achieve this transnational articulation in other ways. The ways in which smaller criminal groups achieve this is through networking with other criminal groups across national boundaries. The networking of criminal groups internationally is now fundamental to the economies of trafficking and smuggling.

Networking between criminal groups can be read as both a pragmatic response to the challenge of achieving and maintaining the transnational mobility of illicit commodities and as an organizational form that offers advantage in the uncertain market conditions characteristic of post-Fordist capitalism within illicit, as well as licit, economies (Harvey, 1989; Hobbs, 1998; Lash and Urry, 1993).

> Though judgements of incidence and prevalence of threats may differ, there is general agreement that networked crime is more efficient than hierarchical "planned centralism" for long-term criminal survival, at both national and transnational levels.
> . . . For criminals intending to stay in business for a long time, unless they are extraordinarily gifted and/or live in an extraordinarily corrupt haven, "small (or at most medium) is beautiful." (Levi, 2002: 906–907)

Externally oriented networked arrangements, for example, offer greater flexibility, agility, and resilience than more centralized, hierarchical forms. As new opportunities and markets open up, or as they are closed down by successful law enforcement or security initiatives, networked arrangements are able to change in response. As Galeotti (2005a: 5) argues, "global organised crime is evolving, embracing new markets and new technologies, and moving from traditional hierarchies towards more flexible, network-based forms of organisation."

Where such networked complementarities can be negotiated and the rapacious impulses to muscle in on another group's turf or market resisted, then it would appear that such alliances can be the strategic outcome. However, these tend to be pragmatic rather than cultural alliances in which the distinct identities of each group within these networked relationships remain largely undisturbed. Galeotti (2005a: 1), for example, describes these relationships as "very loose, little more than mutually-profitable trades or temporary alliances of convenience."

The apparent hegemony of the networked criminal form within transnational illicit markets raises a number of related challenges for security, law enforcement, and policy. While potential responses are explored in detail in Chapter 7, it is worth briefly outlining four of the key challenges here. First, networked criminal forms are flexible and adaptive, certainly more so than more traditional hierarchical, centralized forms. For this reason, they are not vulnerable to decapitation approaches by security and law enforcement agents (the removal or incarceration of criminal bosses, so-called "kingpins" or "Mr Bigs") (Levi, 2014: 12). While the removal of major players within nodes in criminal networks is likely to be disruptive to the groups affected, the flatter organizational structures of these groups suggest it would not necessarily be fatal to them. Unless investigators are able to trace illicit authority throughout a network and achieve decapitation at each node, something that the literature records not at all, such an approach is unlikely to achieve anything more than temporary, localized disruption to the network. International cooperation and coordination between police forces is an increasingly common approach in the face of these networked economies and there are frequent successes that typically draw extensive, favorable media attention (Channel 4, 2010). However, whereas it is typical for these operations to impact on criminal groups at the origin and destination of networks of illicit commodity movement, less is known or said about their impacts on groups involved at the mid- or transit points of these networks unless this transit has particularly violent outcomes (Bhattacharayya, 2005; Brophy, 2008; Glenny, 2008; Watt and Zepeda, 2012). Furthermore, given the sheer size of the illicit trades they are attempting to counter, the global impacts of these successes are limited (Sproat, 2012). There is no evidence that these efforts have fundamentally disrupted these trades.

Even where such successful outcomes are achieved, they highlight the second challenge posed by criminal networks, namely, that they are highly complex in that they bind together multiple places, agents, and activities, each grounded in specific conditions, and that are widely distributed across international space (Murdoch, 1997, 1998; Sheppard, 2002). While the successful arrest of networked groups of people traffickers or counterfeiters in, for example, the United States, the United Kingdom, or China, is to be celebrated, it does nothing to address the conditions within these countries that underpin and sustain the production and consumption of these illicit commodities or the services associated with these networks, not to mention the conditions in spaces through which these commodities move on their journeys, which make the transit of illicit commodities a viable, even attractive, economic strategy for these regions. Thus, it is now widely acknowledged that tackling organized crime requires far more than law enforcement approaches alone (Midgley et al., 2014; Moynagh and Worsley, 2008; Varese, 2011). However, the challenges to enacting alternatives are becoming well documented (Midgley et al., 2014).

Third, both across the wider realm of organized crime generally and even within specific networks, the range of criminal activities is very broad. For example, the economies of counterfeiting involve theft of intellectual property, manufacture of counterfeit goods, possibly using illegal and/or trafficked labor, smuggling, corruption, sale of illicit commodities, money laundering, and possibly violence, all of which demand different policing and policy responses. Yet responses to organized crime have been criticized for their "silo" nature, restricting their focus only to specific aspects of wider networks of criminal activity (Levi, 2014: 11). It is a challenge for law enforcement agents to detect the scale of the networks and activities involved in this one trade alone, let alone devise a range of responses that can tackle all or even most of them in an effective manner. Finally, networked criminal forms are dispersed across international space and this is inherently a challenge to territorially bound police forces (Kirkby and Penna, 2011; Levi, 2014; Madsen, 2009).

In sum, then, the challenges facing international policing, security, and policy communities are achieving consensus around the range of measures that can effectively tackle the scale of the networked criminal economies characteristic of the global illicit economy, rather than achieving individual but ultimately localized successes, and coordinating these measures across international space and the, to date, somewhat discrete realms of policing, security and law, politics, culture, and economic and social development. The limitations and failings of extant approaches to tackling organized crime are reviewed in Chapter 7 and some possible advances outlined there.

REGIONAL DIFFERENCES IN THE ORGANIZATION OF CRIMINAL GROUPS

It is empirically undeniable that there is enormous variation in the organization, nature, and activities of criminal groups around the world (Aning, 2007; Castells, 2000; Davidson, 1997; Glenny, 2008; Hastings, 2009; Hignett, 2005; Hobbs, 2004, 2013; Hobbs and Dunnighan, 1998; Litner, 2005; Lombardo, 1997; Nordstrom, 2007; Scalia, 2010; Stephenson, 2015). However, whereas the literature adequately demonstrates the ways in which criminal organization has evolved temporally, reflecting the broad changes in the nature of capital accumulation discussed above, questions concerning spatial variations in the organization of criminal groups and their markets, reflective perhaps of the geographically variegated nature of capitalism (Peck and Theodore, 2007), global asymmetries (Passas, 2001), the presence of different combinations of locational characteristics in different places, and specific networked relations (Hall 2010a; see also Chapter 5) have gained less purchase within these discussions. Furthermore, there is a lack of consensus around the significance of any such regional differences.

Albini and colleagues (1997: 154), for example, argue that organized criminal activity is reducible to a set of basic characteristics that are reproduced across different regional and international contexts in ways that vary only superficially by location. This universalizing tendency can also be seen in attempts to produce definitions of organized crime through the identification of sets of essential characteristics (Abadinsky, 2013; Albanese, 2005; Finckenauer, 2005). Typically, such definitions speak little, if at all, of geographical variations in criminal organization. However, others have warned of the dangers of universalism in discussions of organized crime, arguing that it differs in its nature, including the ways in which groups are organized, and not just in terms of its extent and the distribution of specific criminal activities, across space:

> It is at the local level that organized crime manifests itself as a tangible process of activity. However, research indicates that there also exist

enormous variations in local crime groups. . . . We found it possible to draw close parallels between local patterns of immigration and emigration, local employment and subsequent work and leisure cultures with variations in organized crime groups. (Hobbs, 1998: 140)

Why, after all, should an organisational model of crime that applies to parts of Italy in some historical periods apply either to the north-eastern US; and even if it accurately depicts crime there, why should it apply throughout, or indeed in any part of the UK, Germany or Canada? (Levi, 2002: 881)

Albini and colleagues (1997) offer a view of the nature of organized crime that stands in contrast to a number of accounts from various disciplinary perspectives including criminology, sociology, political science, and investigative journalism who advance empirical accounts charting the diverse ways in which organized crime differs in its historical evolution, extent, type, market, group organization, and relationship with its social contexts in different regional settings (Allum, 2006; Castells, 2000; Davidson, 1997; Glenny, 2008; Hastings, 2009; Hignett, 2005; Hobbs, 2004, 2013; Hobbs and Dunnighan, 1998; Litner, 2005; Lombardo, 1997; Stephenson, 2015). A cursory reading of these literatures suggests that while geographical differences in criminal organization might be declining (Nordstrom, 2007) as many criminal markets become more networked and outwardly oriented, they are far from having disappeared altogether. Two questions follow here. First, in exactly what ways does the nature of criminal groups vary between regions, and second, in what ways, if at all, are these variations significant? Can they, as Albini and colleagues (1997) assert, be dismissed as superficial or are they worthy of more explicit and sustained attention?

As has already been established, the basis of group membership and subsequently the organizational structures, cultures, and politics of criminal groups vary considerably. We can identify "traditional" organizations based around family ties such as the Sicilian Cosa Nostra, the Calabrian 'Ndranghetta, and some Israeli crime groups (Glenny, 2008: 139; Paoli, 2005); territorial units such as many Russian crime groups (Stephenson, 2015); clan associations such as those observed within Afghani narcotics economies (Goodhand, 2009); or groups based on a common ethnicity, as with Vietnamese cannabis-producing networks (Schoenmakers et al., 2013; Silverstone and Savage, 2010), as well as the many looser criminal associations that are increasingly characteristic of the post-Fordist criminal economies discussed above. The reasons that these associations develop on the bases that they do within different contexts, little studied in any comparative sense, are complex and multiple. They are reflective of combinations of factors such as the natures of the different societies from which these groups have developed, broader economic transformations, their historical continuity or otherwise, their regional economic and political contexts, and the nature of the markets within which the groups are operative (Hastings, 2009; Paoli, 2005; Scalia, 2010). It would appear that no single one of these factors is determinate of criminal organization. Thus, criminal organization might display some regional variation but we cannot argue that it is regionally determined. It is possible, for example, to note the presence of groups organized in different ways within a single region. Paoli (2005: 26), for instance, cites examples of the presence of Italian non-mafia crime groups in Italy alongside its more traditional mafia associations. In being studied primarily through the lens of the individual empirical case study, though, rather than through a more comparative framework, while it is possible to make this observation, it is more difficult to make any authoritative statement about the relative influences of specific factors, including space, potentially influencing the dynamics of criminal organization. Numerous commentators, for example, cite the links between factors such as state weakness and the presence of organized crime within a region (Aning, 2007; Bagley, 2005; Brophy, 2008; Castells, 2000; Goodhand, 2009; Hastings, 2009; Hignett, 2005; van Dijk, 2007), but few go on to link this to the specific nature of criminal groups present there. Presently, perhaps, we can do little more than suggest that criminal organization is reflective of factors either internal or external to groups, or the result of some dialectical combination of these, with reference to individual cases. To do even this, though, would enhance the conceptual framework through which we speak of these groups.

There are occasional exceptions to this situation, however. Hastings (2009) embeds his readings of the nature of groups operating within maritime piracy economies in Southeast Asia and Somalia within their regional economic and political contexts. These, he argues, shape the nature, ambitions, and operation of these groups. Despite operating in the same economy, albeit in different parts of the world, the two regions are characterized by very different types of criminal organization:

> The syndicates themselves (in South East Asia) are also the dark side of the legitimate shipping networks or the shipping side business of

organized crime gangs. Such organizations have neither the means nor the aspiration to wield enormous amounts of fire-power and dominate other groups in the area. By contrast, Somali pirates, as what are effectively the navies of local warlords, have both the desire and opportunity to do so. (Hastings, 2009: 221)

This analysis suggests a profitable line of criminological enquiry in pursuing a more complete understanding of the nature of criminal groups and their multiple regional, economic, social, and political contexts, but the relative paucity of such analysis demonstrates that there is much to be done to produce more comparative understandings of these processes.

The relationships between organized crime groups and their sociocultural and political contexts is important in their being able to forge senses of legitimacy within these settings. This reality has been explored in a number of cases (Castells, 2000; Glenny, 2008; Hill, 2005; Kleemans and van der Bunt, 1999; Stephenson, 2015). Often crime groups are painted as predatory agents seeking to capture elements of their regional contexts for egotistical or pragmatic, materialistic motivations. However, this overlooks the argument that wholly predatory incursions of this kind are unlikely to gain the legitimacy and stability required for them to endure (Kilcullen, 2013). Rather, there is, at times, a symbiotic quality to the relationships between organized criminal groups and their sociocultural and political contexts that often include nonstate actors such as organized criminal groups acting as proxy agents of governance across terrains that remain at least partially closed to conventional forms of state governance, such as borderlands (Goodhand, 2009; van Schendel, 2005), prison systems (Stephenson, 2015: 152), and criminal markets generally. For the communities of regions affected by this political pluralism, such actors can provide, through processes of persuasion, administration, and/or coercion, senses of security and certainty (Felbab-Brown, 2014: 43), and in some cases, employment and income, while extracting resources from these same populations. It should be noted, however, that the social legitimacy that some criminal groups are said to possess has been questioned. Rather, it has been argued, relations between criminal organizations and communities are asymmetrical and coercive rather than organic (Midgley et al., 2014: 27).

While this sociocultural and political embeddedness is important in underpinning and sustaining the economies of organized crime—it was key, for example, in the capture of formerly state resources by organized criminal groups in Russia in the 1990s (Castells, 2000;

Stephenson, 2015)—its role in directly shaping the specific nature of these markets, producing regionally variegated criminal economies, while being anecdotally apparent in the literature, is less certain and has thus far been only the subject of limited substantive or systematic evaluation. Some cases that have been recorded include instances of some criminal groups observing cultural taboos that preclude their opening up or participation in certain markets such as a Taliban prohibition on opium production that existed before 2001 (Farrell and Thorne, 2005) and the prohibition on exploiting prostitution among Sicilian Cosa Nostra and Calabrian 'Ndrangheta mafia associations (Paoli, 2005: 21). Such restrictions, however, can come under strain in the face of compelling market logics. It has also been noted that the strict conditions, not only of membership of these latter groups, which are restricted to men born within the groups' home regions, but also of their areas of residence, has limited the expertise within the groups and negatively impacted their involvement in black markets associated with arms, money, and gold (Paoli, 2005: 23). Typically, the economic logic of criminal markets is an expression of the opportunities available locally or accessible across wider terrains. However, the examples above suggest this logic is not entirely determined economically. Despite this, though, such sociocultural dimensions of criminal markets seem relatively peripheral to their operation in the face of market logics and opportunities. More generally, the articulation of geographical and organizational difference, potentially, suggests the efficacy of more variegated responses to the problems of organized crime. This topic is explored further in Chapter 7.

REGULATING CRIMINAL ECONOMIES

Criminal economies are terrains within which much violence is practiced and acts as a key form of regulatory mechanism. As Hobbs (2013: 171) reminds us, not all of the violence observed in criminal markets is purely instrumental. In some instances, most notably in the case of recent drug-related violence in northern Mexico, this violence is exercised on an industrial scale and is, at least partially, indiscriminate (Gibler, 2011; Watt and Zepeda, 2012; Wright, 2011). However, such extreme levels of violence are not routine, despite what some of the more gratuitous cultural renditions of organized criminal markets might suggest. Violence and the plausible threat of violence deployed through reputation or gossip and rumor (Hobbs, 2013) are key resources, along with forms of trust, that are deployed in the regulation of criminal economies and the maintenance of order therein.

"Regulation" here refers to both the internal regulation of intra-groups' relations and the exchange between different criminal groups within markets and between groups and consumers of illicit products and services. The governance literature has said little about regulatory mechanisms and processes within illegal economies (Hudson, 2014: 785). Clearly, as Hudson suggests, many regulatory mechanisms of licit economies are not available within illegal economies. This does not mean that these markets are unregulated, as such an absence of academic commentary might suggest, however. The lack of formal regulatory mechanisms is significant, though, to spatialized readings of illegal economies because of the roles they, and their associated texts, play in mobilizing social relations of trust beyond copresence and across space. The lack of these mechanisms within illegal economies renders regulation within these markets, potentially, a much more localized, immobile process.

The regulatory mixes within criminal economies are not universal across different markets and territories. For example, trust can be derived from many different reciprocities such as those of clan, ethnicity, kinship, territory, or different combinations of these. These will vary between markets, reflective of their different histories and current circumstances. In some cases, trust, rather than violence, might be the overriding regulatory commodity, as with the cultivation of cannabis in the Netherlands by Vietnamese groups (Schoenmakers et al., 2013). Elsewhere, violence appears to be the, or at least a, primary method of regulation (Brophy, 2008; Saviano, 2016; Vulliamy, 2010; Watt and Zepeda, 2012). However, even in circumstances where violence appears to be the foremost method of regulating illicit economies, it is rarely successfully deployed in isolation (Kilcullen, 2013; Lilyblad, 2014; Wainwright, 2016: 103–105) unless, perhaps, it can be mobilized on the scales that have been observed in parts of Mexico in recent years (Gibler, 2011; Wright, 2011).

Spatially, then, the overriding regulatory mechanisms deployed within criminal economies are grounded and localized in various ways and display only limited mobility. The various associations from which trust in these economies is derived tend to be associations that are, to some extent, in place, if not necessarily solely of place. Associations of family, clan, or ethnicity, for example, tend to be and remain to a large extent contingent within specific territories. Similarly, violence, either in act or through threat, reputation, or rumor, while not impossible, is difficult to deploy at a distance. "It might be taxing to make victims believe that the person standing in front of them belongs to a menacing foreign mafia. A reputation for violence depends on long-term relations, cemented within independent networks of kinship, friendship, and ethnicity. It is next to impossible to reproduce them in a foreign land" (Varese, 2011: 4).

Criminal reputation and rumors are ways in which regulation of illegal markets can be mobilized to some extent. Here, though, this mobilization has tended to be limited to those local or regional social worlds within which particular criminal brands or specific stories have meaning (Hobbs, 2013; Varese, 2011). Having said this, it is not impossible for the regulation of criminal markets to be enacted at distance in various ways. For example, violence and trust can be mobilized through the networks associated with transnational diasporas. However, here their effectiveness is restricted to the spaces of these diaspora communities rather than being more widely applicable in distant spaces. Thus, violence tends to be deployed within diaspora communities in these instances and only rarely spills out beyond them. Violent regulation at a distance is also possible through either the dispatch of violent actors to pursue a distant miscreant or the subcontracting of violent labor to indigenous actors in distance places. However, these methods generate a series of potential risks associated with incursion into territories of alien violent authority, of agreements not being honored in the case of subcontracting violence, or of detection and the "heat" from local state authorities, the media, and indigenous criminal actors that results from a successful "hit" on foreign soil. None of these mechanisms suggest that it is easy to deploy criminal regulation at a distance in any certain or sustained way, and the literature records little evidence of this having been achieved in any systematic and enduring sense (though see the discussion of franchising of the Zetas' criminal brand, discussed below [Wainwright, 2016]).

The problems that criminal actors within cybercrime economies have had in achieving effective forms of regulation further demonstrate the ways in which criminal regulation is a problematically mobilized process. Here the spatially distributed actors within these economies are brought together virtually within secretive cybercrime forums. However, these are only able to provide very weak forms of governance of these markets as the maximum sanction that can be enacted is the threat of reputational damage, difficult in markets characterized by anonymity, and exclusion from these forums. These have proven limited deterrents to aberrant behaviors compared to the violent alternatives available in the local worlds of offline criminal economies (Lusthaus, 2013: 56).

The immobility of informal forms of regulation and their effects on the geographies of markets are also apparent in some licit markets such as the jewelry trade and agriculture. In these cases trust is developed through long-term contact between actors that requires spatial co- or near presence through practices such as physical contact (the "gentleman's handshake") (Fisher, 2013) or the clustering of firms in places such as Birmingham's Jewellery Quarter in the West Midlands of the United Kingdom (CIBJO, 2007; Depropris and Wei, 2007). The difficulties of mobilizing the informal mechanisms that dominate the regulation of illegal economies across transnational space is one of the key reasons that organized criminal groups have remained relatively less mobile than their licit corporate counterparts. As Varese argues, in criminal economies, "distance makes it harder to monitor an employee, and ensure that the agent works efficiently and honestly" (2011: 14).

However, it is clear that criminal actors are aware of the impediment to their expansion that the localized regulatory resources available to them have, and there is some emergent evidence of their attempting to forge organizational forms that are able to more effectively mobilize these resources. Wainwright (2016) discusses the case of the Zetas, a former security unit for Mexico's Gulf cartel which, in recent years, have emerged as a powerful cartel in their own right. The Zetas have undoubtedly grown in recent years and appear to have achieved some significant spatial expansion across eastern Mexico and the Caribbean coast of Central America. This growth, in part, appears to have been facilitated by the Zetas' innovation of a franchise model of organization through the recruitment of prominent criminals in new areas.

> The Zetas have employed a version of franchising. According to analysis by the regional head of the UN's Office for Drugs and Crime, the Zetas have decided not to send their own representatives to new markets to set up criminal outposts from scratch and instead have adopted local gangsters into their club, as franchisees. . . . The Zetas' central command provides the franchisees with military training, and in some cases arms. In return, franchisees share a slice of their revenues with the central organization and agree to a form of "solidarity pact," an agreement that they will fight for the Zetas if a war breaks out with another cartel. (Wainwright, 2016: 153)

This is the only incidence of franchising recorded in the literature thus far, and it remains to be seen how enduring and resilient these arrangements prove. Little specific analysis has been conducted on franchising within criminal organizations and it is unclear as to the extent to which this represents a new model of criminal organization and mobilization or a possible development of the existing networked arrangements outlined above. The expansion of the Zetas, in this way, has been observed across a contiguous region. It is not known, as yet, to what extent these arrangements might facilitate international criminal mobilities. This example, though, does provide a rare insight into the mechanics of the arrangements between criminal groups within networks.

There has been little specific empirical investigation and theoretical reflection upon the precise mechanisms of regulation that emerge and operate between different criminal groups within plural, diverse illicit networks that are dispersed across transnational space (though see some studies cited in von Lampe, 2012: 193). Further research into the regulatory mechanisms and trajectories of cases such as these would undoubtedly advance our understanding of networked, and potentially franchised, criminal forms and the regulation of the illegal markets and the spatialities of these markets.

Summary

This chapter has scanned a broad literature on criminal organization, drawing particularly heavily, but not exclusively, on that from a criminological perspective, which has produced case studies of criminal organizations. A number of key points emerge from this endeavor. We can, for example, see that the contemporary landscape of organized crime is characterized by a diversity of organizational forms that vary in terms of their size, structures, the markets they operate within, the regulatory resources deployed, and the bases upon which these groups are organized. The question of criminal organization, though, cannot be reduced to being the product of economic rationality. This chapter has shown how criminal organizations are both economic and social units. While this is true of all enterprises, intuitively at least, it seems that the social has a more determinate influence over the nature of criminal organizations than it does over enterprises operative within licit economies. Some, for example, attribute flatter organizational forms that have been observed recently to the social relations that exist between members of some criminal groups

(Hallsworth, 2013; Stephenson, 2015: 105). To extend this claim across licit economic terrains, however, is an assertion that is based on little empirical evidence. The question of the relative influences of the social dimensions of enterprises on their nature and organization within licit and illicit economic realms is one that would repay further investigation and reflection.

Criminal organization is also dynamic and is reflective of wider economic changes. This chapter has recorded examples of criminal organizations innovating flexibilities within their organizational forms through the emergence and spread of less rigidly organized associations and networked arrangements between discrete groups, and the possible recent emergence of franchising arrangements in certain criminal markets. None of these observations should necessarily surprise us, given the wealth of literature that now records them. We are left, though, with the question of to what extent these changes represent fundamental or partial transformations of criminal enterprises and their organization. The evidence reviewed in this chapter conveys senses of empirical changes within the markets and groups discussed. However, this question has not been yet perused in any explicit, comprehensive, and systematic sense. Undoubtedly, whatever tentative conclusions have been sketched here, much remains to be investigated. Collectively, it would seem that criminal economies are now more plural terrains, in terms of the groups that populate then, than they were at any point historically. There is also ample evidence that individual groups are less exclusively defined through the ethnicities of their members. Ethnicity, however, cannot be discounted as a factor in shaping criminal organization. Similarly, the significance of place and the rootedness of criminal organizations persist in many instances and mobilizing beyond their home region remains a challenge for criminal organizations. Geography, then, remains an influence on the nature and organization of criminal economies in multiple ways. Networked complementarities allow organized criminal groups to articulate with extensive transnational economies. These networked arrangements appear to be more collaborative than integrative in nature, but the possible emergence of more franchised associations does raise questions about whether we can think about these relationships in such binary terms. Spotting and scrutinizing trends in the relationships between criminal groups within and across markets, then, should remain a priority for those investigating criminal groups and economies, whether they do so from policing and security or from academic perspectives.

CHAPTER 5

THE SPATIALITIES OF ORGANIZED CRIME

INTRODUCTION

This chapter considers the spatialities of organized crime, their extant geographies. "Geographies" here encompasses, as well as the spatial distributions of organized criminal activities and examples discussed in Chapter 2, fundamental questions of scale, or rather scales. There is no natural or obvious scale at which we should observe the economies of organized crime. Political and policy discourses, for example, typically reference organized crime through the global, regional, and national scales. However, these are socially constructed lenses reflective of the rhetorical elision, within these circles, of "organized" with "global" crime (Hobbs, 2013: 149–150) and the imagination of its threat to national security (Sproat, 2012: 328); the equation of certain world regions with crime problems, for example, Latin America with the production of an international drug problem impacting on the United States, often as a mask to further other economic and political agendas (Gregory, 2011: 79; Watt and Zepeda, 2012: 204–206); and the nation state as the primary spatial unit within which many anti-organized crime responses are formulated and resources marshalled (Hudson, 2014: 789). However, turning to filmic discourses, it is the local scale that emerges most sharply. For example, the microgeographies of New Jersey's commercial, industrial, and suburban landscapes unfolded across the opening titles of that great American television drama *The Sopranos* (1999–2007). Its six seasons were rooted, save for a few excursions to places beyond, in these very settings. Likewise, the claustrophobic atmosphere in director Matteo Garone's *Gomorrah* (2008), a film dramatization of Roberto Saviano's (2008) undercover exposé of the Neapolitan Camorra, discussed in Chapter 3, was achieved in part through its depiction of a series of grim local geographies. The specific localities of these dramas, though, were also articulated through their connections to other spaces. Stolen and counterfeit goods and narcotics arrived in shipping containers from China and elsewhere to be distributed to their local markets. Disputes with Russians and Poles unfolded.

All of these scales of analysis have their merits and all have been embraced within the multidisciplinary academic literatures of organized crime. All reveal different aspects of these illegal economies. Equally, however, each, on their own, conceal aspects that are only apparent at other scales. Thus, developing a sophisticated understanding of organized crime demands that it is looked at across multiple scales. This chapter considers the factors underpinning organized criminal activities at the global, regional, national, and subnational scales. It would be possible to extend this somewhat further than is done here though through the incorporation, for example, of feminist literatures that emphasize more the scales of the body and the household. These have proven influential generally with economic geography (Marston and Smith, 2001) and have enjoyed some limited, if potentially growing, purchase within the literatures of organized crime and illicit activities (Massaro, 2015; Meehan, 2012; Wright, 2011).

Economic geography takes on organized crime could adopt, as outlined in Chapter 1, many different approaches. This chapter aims to do more than simply outline a litany of factors and scales that might explain specific geographies of illicit and illegal economies. Rather, it will, for example, seek to problematize the scales of analysis at which it speaks, highlighting their contingencies. It will try to show how multiple scales mutually constitute each other simultaneously. This chapter can be seen, in part, as a response to the point outlined in Chapter 1 that geographers have failed to say much about organized crime specifically and illicit economic activities generally, despite the wealth of insights a geographical perspective might provide (Brooks, 2012: 3; Hall, 2012, 2013; Hudson, 2014, 2016).

The chapter is organized around a number of points. The first is simply the empirical reality that organized criminal activities are distributed unevenly across global space. Like all economic activities, they display distinctive geographies, many of which were introduced in Chapter 2. These geographical patterns are reflective of processes operating and factors present at a number of scales. This book contends that these geographies have not been taken seriously enough within the multidisciplinary extant literatures of organized crime, nor has organized crime been engaged with seriously enough, with a few exceptions (Allen, 2005; Brown and Cloke, 2004, 2007; Hall, 2013; Hudson, 2014; Rengert, 1996) within the literatures of geography. This chapter goes in search of those many factors that have been recognized as shaping the global, regional, national, and local geographies of organized crime. The chapter then moves on to two sections that reflect on the value of the knowledges discussed here. The first explores the potentials of analysis that recognizes multiple rather than single scales and the contributions of network ontologies to complementing these perspectives. Emerging within sociology and while subsequently spreading across many fields of social research, finding a particular purchase within economic geography (Dicken et al., 2001; Murdoch, 1997, 1998; Sheppard, 2002), network ontologies have had only limited impacts within criminology generally and the literatures of transnational organized crime specifically. They seek to transcend many enduring dualisms, such as local–global, and offer potential alternatives to the locational, scalar approaches typical of the social sciences and which have been influential within discussions of transnational organized crime. Finally, the chapter outlines some tentative observations on how these knowledges might be applied to the problem of formulating measures and policies for tackling organized criminal economies. These are issues that are returned to in greater detail in Chapter 7, however. The chapter, then, attempts to advance a perspective that is multiple, networked, and applied.

GLOBAL ECONOMIC CONTEXTS

Drawing on insights from institutional and evolutionary political economy perspectives (Hudson, 2005: 13), there has emerged a broad agreement across an extensive critical literature that the neoliberal global economy as it has recently evolved is facilitative of illicit economic activities generally, and those of organized criminal groups specifically, in a variety of ways. These relate to both general conditions within this realm and the specific opportunities it generates (Aas, 2007; Albanese, 2005; Bhattacharyya, 2005; Castells, 2000;

Hudson, 2014; Midgley et al., 2014: 10, 35–40; Moynagh and Worsley, 2008; Nordstrom, 2007; Passas, 2001). These, as we saw in the previous chapter, in turn have had a transformative effect on the economies of organized crime, making them more plural, international terrains (Aas, 2007; Abraham and van Schendel, 2005; Madsen, 2009; Weinstein, 2008).

As we have seen, the assumption that states stand in categorical opposition to illegal economies is one that cannot be sustained empirically or conceptually across a number of historical and contemporary terrains. Some writers, for example, have explored the roles of organized criminal groups within major historical–political processes that have underpinned the emergence of the contemporary global economic system. Historian Michael Woodiwiss (2001; see also Andreas, 2014; Dickson-Gilmore and Woodiwiss, 2008: 68; Hobbs and Antonopoulos, 2013: 31; Woodiwiss, 2012; Woodiwiss and Hobbs, 2009) has explored the roles that organized criminality has played in the emergence of a modern America, highlighting examples of both collaboration and confrontation between government and organized criminal groups. Some writers, extending this theme conceptually, have argued that the illicit is an inherent quality of seemingly licit political and economic systems and have drawn on the pioneering work of Smith (1980) and Tilly (1985), who have proposed, respectively, a continuum of economic enterprises that span the licit–illicit divide and continuums of state practice and war making that encompasses organized criminality (Wilson, 2009: 44; Wright, 2006: 56). Other writers speak of specific historical moments where, while it is clear that illicit actors have benefited from developments and transitions within economic and political systems, they may have also played key facilitating roles within these transitions. This has been clearly observed in the chaotic, yet fundamental, transitions to capitalist market economies across post-Soviet space (Aas, 2007: 12; Bhattacharayya, 2005: 74; Castells, 2000; Stephenson, 2015; Swain et al., 2010). Aas (2007: 11) also discusses recent developments such as the roles of organized criminal groups in weakening state sovereignty and deregulating national economies through the privatization of state assets in transitional economies.

Furthermore, the literature is replete with examples of licit actors who benefit from the activities of organized criminal groups (van Schendel, 2005: 60–61; Weinstein, 2008). These include links between illicit and illegal activities such as narcotics economies, counterfeiting, and money laundering in the large cities of the global

South (Bhattacharyya, 2005: 99; Daniels, 2004: 506–507); the benefits of illicit financial flows to mainstream financial institutions (Brown and Cloke, 2007; Madsen, 2009: 110); and benefits to market economies from illegal people smuggling (Aas, 2007: 40). As was outlined in Chapter 1, it has been observed that where the material benefits of illegal activities outweigh those available in licit economies, coincide with significant consumer demand, and/or where illegality is seen as one of the few ways for economies to access circuits of international capital (Hudson, 2014: 786) is not unknown to see social and, on some occasions, (tacit) state toleration of organized criminality (Bhattacharyya, 2005: 98; Hornsby and Hobbs, 2007). Thus, while such activities are formally illicit, in certain contexts they may be seen as socially licit (Abraham and van Schendel, 2005; Chiodelli et al., 2017; Nordstrom, 2010: 173). Such illegal economic activities, it has been argued, may constitute nationally significant, if only partially glimpsed, economic mobilities. "Countries may formally decry extra-legal practices, and some may be genuine in their denunciations. But in terms of a nation's sheer monetary bottom line, such activities bring in cash that girds the viability of the state. These activities can yield hundreds of millions of dollars yearly for a single country" (Nordstrom, 2010: 173).

Regulatory Asymmetry, Mobility, and Opacity

A host of regulatory and institutional contingencies open up the spaces within which organized crime and other illicit economies may thrive (Midgley et al., 2014: 35). Despite political discourses commonly stressing transparency and regulation, the global economy is a realm defined equally by opacity and anonymity. A number of major recent and contemporary geopolitical changes, for example, have had the effect of undermining the capacities to govern and regulate certain regions while generating or accelerating the development of international networks that are either inherently illicit in nature or that are open to appropriation or use by actors operating within illicit economies. These geopolitical changes include the economic and political transformations of post-Soviet space, the emergence of capitalism in China, and processes of planetary urbanization that have seen the increasingly rapid growth of megacities in the fragile littoral zones of the global South (Castells, 2000; Glenny, 2008; Hudson, 2014: 781; Kilcullen, 2013). Thus, the issue of regulation, or rather its evasion, is a key factor underpinning the spatialities of illicit and illegal economic activities specifically and liminal economic activities generally (Zook, 2003). Illicit economic activities thrive in spaces of opacity, anonymity, and ambiguity where the gazes of the state and international authorities and law enforcement agencies are uncertain or absent altogether.

In part, the inherent regulatory asymmetries and opacity that are now so characteristic of the global economy derive from its more general recent historical development, which, since at least the 1970s, has been characterized by tensions between regulatory landscapes overwhelmingly articulated through the nation state, or in a few cases multinational territorial units, and economic processes and their associated geographies that are increasingly transnational and mobile in nature. As Leyshon (1992: 251) has argued, "the globalization of markets and firms has served to undermine the coercive power of regulatory systems which are embedded within particular geographical–political jurisdictions." A more recent assessment by Midgley and colleagues, drawing on the United Nations Office on Drugs and Crime, affirms that little fundamentally has changed in the intervening 22 years since Leyshon's comments. "The process of globalization has outpaced the growth of mechanisms for global governance, and this deficiency has produced just the sort of regulation vacuum in which TOC (transnational organized crime) can thrive" (2014: 9). While regulatory environments, be they legal or social, are territorially bound, the goods and activities they seek to regulate are increasingly mobile and connected. Licit- and illicitness, then, is rendered contingent and provisional upon the interplay of regulation, space, movement, and connection.

Opacity is produced, for example, through the complex and extensive contemporary mobilities of globalization, which have proven impossible to adequately monitor and secure. The modest standardized shipping container, developed in the mid-20th century, has proved to be a revolutionary technology of globalization, facilitating the easy mass movement of goods at low cost that "redrew the world's economic geography" (Urry, 2014: 33), allowing the global redistribution of manufacturing (see also Martin, 2015, 2016). Ninety percent of the world's cargo now travels by container ship on routes running predominantly from the global South to the global North (Glenny, 2008: 389; Nordstrom, 2007: 115; Urry, 2014: 33–34). Thus, Urry can argue "container ships realised the greatest ever movement of material objects in human history" (2014: 34). These technologies are particularly attractive to criminal entrepreneurs for the same economic and logistical reasons that they work in licit economies. Namely, they provide cheap, easy, secure, and relatively quick ways of moving goods globally. It is no surprise, therefore, to discover that illicit and illegal commodities of all types are routinely hidden, typically among legal cargoes, within shipping containers (Nordstrom, 2007: 115; Varese, 2011: 175).

The success of this system of commodity mobility has generated its own, now inherent, opacity, however. The volume of shipping containers continually entering and leaving ports, the complexity and time involved inspecting them (an average of 5 hours per container), and the economic pressures to move containers through ports, means only a tiny proportion can ever be checked. While the most well-equipped ports in the world can inspect up to 5% of the containers they receive, the reality for many smaller or less sophisticated ports is more like 1% (Nordstrom, 2007: 118, 160; Midgley et al., 2014: 9). The regulatory surveillance of this trade, then, is largely undermined by its volume and the imperatives of continuous mobility. Nordstrom outlines the daily challenges for Europe's largest port, Rotterdam:

> In the Port of Rotterdam's success, size and sophistication lies the secret of the extra-legal. . . . On an average day, 81 seagoing and 364 inland vessels visit Rotterdam, carrying 1.095 million metric tons of cargo and 24, 657 containers. Only the smallest percentage can be checked by police, security, or customs. Rotterdam has portable state-of-the-art scanning machines. While this sounds like an ideal solution, it takes several hours to scan a single container. . . . Large, 6, 000-plus container ships are common here: consider inspecting a container at the bottom of the ship under tightly packed, towering stacks of multi-ton metal boxes that can only be moved one by one. (Nordstrom, 2007: 160)

Container security initiatives that stress transparency are susceptible to compromise through technological limitations and corrupt practices embedded within the infrastructures of global trade. The Container Security Initiative (CSI) was introduced by the United States in 2002 in partnership with a number of foreign governments (Glenny, 2008: 389). By 2014 58 ports outside the United States had signed up to the initiative, which involves prescreening containers in foreign ports before they are shipped to the United States, after which they are granted free entry (Urry, 2014: 152; Varese, 2011: 177). The initiative involves stationing U.S. customs officers in overseas ports. However, Glenny has pointed out that the effectiveness of the initiative depends on the experience and linguistic skills of these officers. In reality, few of the officers stationed abroad under the CSI had previously worked outside the United States and fewer still had adequate language skills, especially in places like Hong Kong and Taiwan, ports where containers originating in North Korea and destined for the United States, pass. To overcome these limitations, U.S. customs officers have relied upon local customs officials who have been recognized as particularly susceptible to corruption and bribery (Glenny, 2008: 390) and who direct the searches that U.S. officials undertake (Varese, 2011: 177). Furthermore, technological limitations mean that while nuclear and radiological materials can be detected, other illicit and illegal commodities are not picked up by scans (Varese, 2011: 177). Two authors who have examined the CSI have concluded that the illusion of transparency it promotes actually facilitates the illicit trades it is designed to prevent (Glenny, 2008: 389; Varese, 2011: 177). It should be noted, though, that the opacity of the shipping container appears to have been exploited with alacrity by actors from within the licit economy as well as criminal entrepreneurs. It is thought, for example, that up to 20% of all goods moved by licit corporations are undeclared in attempts to avoid taxes and duty (Nordstrom, 2007: 172). Shipping containers, then, offer opaque microspaces that are readily available to criminal and other illicit and liminal entrepreneurs.

A host of other similarly opaque microspaces are utilized for the production, storage, and distribution of illicit and illegal commodities. These include laboratories used for producing, synthesizing, and refining narcotics (United Nations Office on Drugs and Crime, 2014b: 107); apartments and houses used to manufacture counterfeit cigarettes, store smuggled and stolen goods, or grow cannabis (Hornsby and Hobbs, 2007; Silverstone and Savage, 2010: 29); and warehouses and industrial units used to store a host of trafficked goods (Hobbs, 2013). In some cases, the opacity of these spaces might be produced through their location in regions geographically remote from centers of state authority. This has been the case with many "*cartelitos*," small narcotic production and trafficking organizations that have grown in importance within drug economies, most notably in Colombia since the demise of larger cartels (Castells, 2000; Kenney, 2007: 259; Saviano, 2016: 159). Many of these have exploited the institutional limitations of authority found within medium-sized and small Colombian cities through bribery and intimidation, and used this to throw cloaks of anonymity around their activities (Bagley, 2005: 39).

Geographical remoteness or mobility, though, are not necessary prerequisites to the maintenance of anonymity in narcotics production and trafficking networks and other illicit enterprises, however. Many units, for example, involved in the synthesising of narcotics, or the growing of cannabis, are found close to their markets in cities of the global North (Abdullah et al., 2014: 14; Silverstone and Savage, 2010; United Nations Office on Drugs and Crime, 2014b). Similarly, the trucks, vans, apartments, houses, warehouses, and industrial units utilized by criminal entrepreneurs are often metropolitan in location (Hobbs, 2013). Here anonymity is, in part at least, sustained through regulatory asymmetries across a number of commercial sectors such as transport, distribution, and haulage (Bucquoye, Verpoest, Defruytier, and Vander Beken, 2005; Europol, 2007: 11; Kilma, 2011), commercial property, privately rented residential accommodation (Nelen, 2008), and cash-rich retail and commercial enterprises (Ferragut, 2012: 15). These sectors are often characterized by combinations of poor or unstable market conditions (Kilma, 2011: 205); a low level of technology (Lavezzi, 2008: 202); and the presence of a diversity of actors from across the licit–illicit continuum, including small companies, the self-employed, those seeking secondary or subsidiary incomes, and those combining licit and illicit business practices such as evading taxes and duties or laundering illicit finance through otherwise legitimate front businesses. Here the penetration of formal regulation is often incomplete (Kilma, 2011: 211). This contingent regulation is supplemented by systems of informal regulation exercised, to some extent at least, through trust, intimidation, and the implicit or explicit threats of violence (Hobbs, 2013; Williams, 2004). The regulation of such markets was discussed in more detail in the previous chapter.

Offshore Tax Havens/Financial Centers, Illicit Finance, and Development

Globally, the most significant forms of opacity are the low and no-tax regimes and secret banking services offered by offshore tax havens. Although tiny in spatial extent, these microstates underpin an extensive offshore financial system that lies, largely, beyond the gaze of state institutions (Denault, 2007; Hampton and Levi, 1999; Hudson, 2000, 2014; Nordstrom, 2007; Roberts, 1995; Sharman, 2011; Stewart, 2012; Urry, 2014). It is easy, given their size, to imagine that these spaces might be marginal to the shape and processes of the contemporary global economy. This would be far from accurate, however. Urry (2014: 27; see also Foltz et al., 2008; Hampton and Levi, 1999: 132; Hudson, 2014: 787–788) places the offshore tax haven at the heart of contemporary global capitalism.

> Core then to neoliberalism has been the large growth in the movement of finance and wealth into and through the world's sixty to seventy tax havens. . . . The growth of what are also known as "secrecy jurisdictions," or in France as *"paradis fiscal,"* are central to the neoliberalisation of the world's economy since around 1980. (Urry, 2014: 46)

Historically, the extensive offshoring of finance by corporations, the superrich, corrupt politicians, and criminal entrepreneurs is a recent, but rapidly growing, phenomenon (*The Economist*, 2013c). In 1968, roughly US$11 billion was offshored through these financial centers. This rose to US$385 billion 10 years later, to US$1 trillion by 1991, and to US$21 trillion by 2010 (Urry, 2014: 46–47). Evidence suggests that 18% of global GDP equivalent is held in offshore funds (Dick, 2009: 98), while 40% of world trade in goods and 65% in hard currency is processed through offshore tax havens (Nordstrom, 2007: 65, 172) along with 50% of international bank lending and 33% of foreign direct investment (Hudson, 2014: 787).

Money that is moved offshore effectively disappears from the state in that it is not available to governments through taxation. Typically, extant discussions about national, regional, and global development do not include reference to offshored finance. The offshoring of finance, though, is not something that takes place solely on the digital screens of banks and other financial institutions. It is now recognized as having had profound impacts on the fates of individual nations, particularly in the global South; on regional and global development over the last 30–40 years; and on global measures of inequality.

> The world's poorest countries, particularly in sub-Saharan Africa, have fought long and hard in recent years to receive debt forgiveness from the international community. . . . In many cases, if they had been able to draw their richest citizens into the tax net, they could have avoided being dropped into indebtedness in the first place. (Stewart, 2012: 38)

The burden of offshoring clearly falls heavily upon the nations of the global South. Research, for example, for the NGO Global Financial Integrity suggests that illicit financial flows from the global

South increased from US$297.4 billion in 2003 to US$991.2 billion in 2012 (Kar and Spanjers, 2014: vii). Over this period, this illicit finance leaving the global South was roughly equal to that flowing into the region from official development assistance and foreign direct investment (Kar and Spanjers, 2014: 12). The countries of sub-Saharan Africa, which include some of the poorest in the world, lost US$528.9 billion in illicit capital flight offshore between these years, which represented 5.5% of the GDP of the region (Kar and Spanjers, 2014: 7, 12).

This suggests that focusing discussions of economics and global development only on the licit/visible aspects of the world economy is a misleading and potentially damaging endeavor (Nordstrom, 2011: 13). Prevailing political narratives around financial offshoring tend to paint a picture of lightly governed rogue microstates playing fast and loose with regulation while they chase forms of mobile international illicit finance in attempts to compensate for their own indigenous economic lack (*The Economist*, 2013a: 1). This is a misrepresentation on four counts. First, tax havens are not restricted to tropical islands such as the Seychelles and the Cayman Islands. Rather, they are also found close to, linked to, and even at the centers of major Northern OECD economies. Tax havens include, for example, Switzerland and Luxembourg in continental Europe; the British Crown Dependencies of Jersey and the Isle of Man—indeed, roughly half of all tax havens are connected to the United Kingdom in some way (Hudson, 2014: 787); and the U.S. state of Delaware and the U.S. territory of Puerto Rico. Furthermore, some global cities such as London, Sydney, and New York offer anonymous investment opportunities to overseas investors that have made them attractive potential or actual destinations for terrorist finance and money laundering (Hudson, 2014: 787; Roberts, 1999: 132; Unger and Rawlings, 2008: 332, 335, 349). There are, then, regulatory and fiscal asymmetries at the heart of these OECD economies (Cameron, 2008: 1151). Second, the development of tax havens was encouraged by the British government until the late 1990s as a route to development for some of its former colonies where few, if any, alternatives existed (Hampton and Levi, 1999: 649). Once established, other islands not connected to the United Kingdom recognized this prospect as a potentially attractive development path and in some cases followed suit. It would be wrong, therefore, to paint the offshore economy as solely the creation of a bunch of aberrant, rogue microstates (*The Economist*, 2013a: 3; Hudson, 2014: 788). Rather, this economy derives largely from the actions of OECD economies and the international institutions dominated by them (Hudson, 2014: 787; Sharman, 2011: 89–90; Urry, 2014). Third, it has been argued that financial opacity is a problem, perhaps to a greater extent, with the economies of major OECD nations than it is within tax havens and offshore financial centers. Sharman (2011: 87; see also Cobham et al., 2015), in an empirical test of the effectiveness of anti-money laundering legislation, found that prohibitions on anonymity were frequently less well enforced in OECD economies than they were in tax haven microstates. This was confirmed in a recent special report into offshore finance for *The Economist* journal:

> Small OFCs (offshore financial centres) and developing countries have been arm twisted into adopting a stringent set of rules, which they have done mostly without kicking up a fuss for fear of being blacklisted. Meanwhile the advanced economies that insisted on the standards, and to which they are probably better suited, have been less conscientious in applying them. The majority of OECD countries are only partially compliant with the FATF [Financial Action Task Force] standard on effective sanctions against failures of anti-money-laundering measures. (2013b: 7)

Finally, tax havens, although they do attract a great deal of criminal, terrorist, and corrupt finance, could hardly be said to rely solely only on these sources for their survival, as we have seen earlier. The majority, roughly two-thirds, of finance that moves through these tax havens derives from actors conventionally located within the licit economy, for whom the avoidance of tax and duty is routine corporate practice (Hudson, 2014: 788; Urry, 2014).

These offshore spaces exemplify the difficulties of separating the licit and illicit, empirically and ontologically, within a variety of contexts. This is axiomatic within the critical literatures of organized crime (Ass, 2007: 125; Bhattacharyya, 2005: 63–64; Hudson, 2014; Nordstrom, 2007; Wilson, 2009; Wright, 2006: 52). It is in these spaces that "dirty," "clean," and "gray" money mingle through a multitude of complex, opaque corporate forms and financial products. The designation of legality and illegality here depends hardly at all on empirical practice—for example, (illegal) money laundering and (legal) tax avoidance share many characteristics—but rather through the normative identities of the actors behind them.

We have, then, a global economic landscape that is asymmetric, opaque, ambiguous, mobile, unstable, and incomplete. It is the combination and the interplay between these qualities that renders it so amenable to those, both legitimate and criminal actors, and those who blur this distinction, who do not wish to make their economic activities visible and transparent.

LOCAL GEOGRAPHICAL CONTEXTS

Structural accounts of the economic geographies of the licit global economy make much of the multiple asymmetries that unfold across global space, which are being intensified under contemporary conditions of globalization. In this broad sense, there is much similarity between the macroeconomic geographies of the licit, illicit, and illegal global economies. Indeed, if we return to the case of commodity movement in shipping containers, then they (literally) intermingle in the same microspaces and global logistics and are articulated by the same macroeconomic logics. Passas (2001) has recognized inequalities across global space as the generators of criminal mobilities running predominantly from the global South to the global North. These criminogenic asymmetries, defined as "structural disjunctures, mismatches and inequalities in the spheres of politics, culture, the economy and the law" (Passas, 2001: 23), underpin the movement of people and commodities that are either illicit or illegal in themselves or are moved in ways or through spaces that render them thus (Aas, 2007: 2). While plausible at a global or regional scale, Passas's concept of criminogenic asymmetries seems to have less to say about these movements at the microscale. Here the complex patterns of criminal activities may respond more to local factors and are less likely to be reducible to an overriding structural logic (Aas, 2007: 125; Abraham and van Schendel, 2005: 4).

Daudelin's (2010) account of drug-related violence in the Americas, for example, is explicitly critical of macroexplanations of organized crime, especially those citing inequality as a key explanatory variable. Rather, Daudelin outlines patterns of drug-related homicides that display marked local and shifting contingencies. These, it is argued, reflect changes over the short term in relatively small localities, the regulation of local drug markets, unstable competitive relations between traffickers, and the geographies of enforcement efforts and their impacts (Daudelin, 2010: 5–8). Picking up on the importance of the local, the United Nations Centre for International Crime Prevention (2002: 14, cited in Hobbs, 2013: 222) argue, "Analyzing organized criminal groups outside of their cultural and social context runs the danger of attributing broadly similar causes for their development in any society, and while these may be accurate, ignores important local causal and contextual issues." Passas's account, then, while valuable, seems to highlight the limitations of perspectives that operate largely at a single scale of analysis generally and at the macroscale specifically.

There is ample evidence in the literature that the geographies of organized crime are not reducible solely to such macrostructural explanations. This would be to overlook the considerable national, regional, and local variations in organized criminal activities that have been observed empirically. While global economic conditions might broadly frame these geographies, there are clearly factors operating at a variety of other scales that shape their manifestations more locally. A counterpoint to work at the macroscale has been the emergence of bodies of sophisticated, empirically rich analysis of the complex local terrains of transnational organized crime (Allum, 2006; Hobbs, 1998, 2004, 2013; Hobbs and Dunnighan, 1998; Stephenson, 2015). This work, moves beyond a simple concern with mapping the spatial distribution of organized criminal activity and seeks to excavate the complex mosaics of local differentiation within broader patterns. Thus, here we see place, the social, economic, cultural and political characteristics of space, emerge as a key variable in the explanation of the geographic variations in criminal economies and cultures. Hobbs, for example, has recognized the social makeup of neighborhoods, local housing policy, and the organization of licit economies as key determinates of the nature of organized crime in different localities.

> It is at the local level that organised crime manifests itself as a tangible process of activity. However, research indicates that there also exist enormous variations in local crime groups. In this report we found it possible to draw close parallels between local patterns of immigration and emigration, local employment and subsequent work and leisure cultures with variations in organised crime groups. (Hobbs, 1998: 140)

This work certainly brings to the literatures of organized crime a geographical inflection and level of detail absent from many accounts constructed at wider scales. However, somewhat disappointingly, they remain atypical of analysis of organized criminality, which still seems to reify more the national, regional, and global scales.

Authors whose work acknowledges the local scale to varying degrees have either argued that the geographical specificity of organized crime is important to nuanced readings of it (Aas, 2007: 174–175; Allum, 2006; Hobbs, 2013; Levi, 2002: 881; Wright, 2006: 5) or have revealed this specificity empirically. Examples of the latter include discussions of the unequal distribution of organized criminal groups across national space (Calderoni, 2011); differences in the nature of these groups in different territories (Cressey, 1997; Hignett, 2005; Hill, 2005; Scalia, 2010); geographical concentrations of specific criminal activities such as cocaine production (Allen, 2005; Bagley, 2005; Castells, 2000; Kenney, 2007), opium production (Goodhand, 2009; Urry, 2014: 164), cannabis cultivation (Glenny, 2008; Schoenmakers et al., 2013), cybercrime (Glenny, 2008, 2011; Madsen, 2009: 3), maritime piracy (Hastings, 2009; Pham, 2011), and ATM skimming (Glenny, 2011: 204); and the highly uneven international migration of mafias (Varese, 2011).

Much work has explored the nature of organized crime through the format of the national or regional case study (see, e.g., Aning, 2007; Bagley, 2005; Brophy, 2008; Castells, 2000; Hill, 2005; Slade, 2007; and regular contributions to the journals *Global Crime* and *Trends in Organized Crime*). Such case studies provide rich narratives highlighting local contingencies in the nature of organized crime in different places, its histories, origins and underlying locational causes. What limits such case studies, though, is that in themselves they contain little or no comparative dimension, which is understandable given their primary focus on the local. To date, there has been little endeavor to draw together such regional case studies, pool their analyses, and seek out general explanations for the development of organized crime within different regions. Put simply, can we, from this literature, identify a set of location factors that are prerequisites for the development of extensive illegal economies within nations and regions within which organized criminal groups act as key market players?

Analysis of a selection of this regional literature suggests that there are a number of factors that occur regularly (see Table 5.1) within these accounts. These, then, are our starting points in determining whether the development of organized crime is reducible to the presence of a series of location factors. It should be noted that there are some contradictions within these factors. For example, a number of authors have noted economic contraction as a factor in the development of criminal economies in places (Aning, 2007; Castells, 2000; Glenny, 2008), while others have noted opportunities for criminal protection in rapidly expanding, if poorly regulated, markets (Varese, 2011).

> **TABLE 5.1. Location Factors for the Development of Organized Crime**
>
> - State weakness/weak rule of law
> - Institutional corruption/close links between state institutions and organized crime
> - Geopolitical/economic transitions
> - Contraction in licit economy/high unemployment/lack of legitimate economic opportunities
> - Demand for criminal protection in expanding markets
> - Strategic location (typically on trafficking routes)
> - Unprotected borders/poor border security
> - Cultural idolization of gangsters/normative influence of organized criminality over cultural realm
> - Social acceptance of illicit and/or illegal economic activities/cultures of illegality
> - Skepticism toward/lack of attachment of population to the state
> - Existence of criminal traditions
> - Inhospitable/remote terrain
> - Presence of foreign mobsters
> - Access to large numbers of weapons
> - Technological advances
> - Prohibition policies
> - Cultural traditions exploited by organized criminal groups
> - Rootedness of criminal groups to place
> - Traditions of violence
> - Appropriate environmental characteristics (e.g., for the cultivation of narcotics)
>
> *Note.* Derived from an analysis of 17 regional case studies. Order reflects the frequency with which factors occurred within accounts.

State Weakness and Organized Crime

It is axiomatic within the literature that weaknesses and contingencies in state authority and the rule of law open up the spaces within which organized crime can become extensive and embedded (Abraham and van Schendel, 2005; Castells, 2000; Costa and Schulmeister, 2005; van Dijk, 2007; Hastings, 2009; Hignett, 2005; Kilcullen, 2013; Kupatadze, 2007; Levi, 2014; Lilyblad, 2014; Midgely et al., 2014; Varese, 2011; Wright, 2006). Contemporary state weaknesses can exist at a variety of scales and for many different reasons. It is rare, although not unknown, for states to weaken or fail to the extent that they are unable to control organized crime within their territories (Levi, 2014: 11). These state weaknesses might be internally generated, namely, an inability of official authority to extend governance and the rule of law across their territories, for various reasons (Abraham and van Schendel, 2005; Aning, 2007; Brophy, 2008; Costa and Schulmeister, 2005; Kupatadze, 2007), or externally generated where state capacity is undermined by economic and political developments at the transnational level including those associated with Northern interventions (Goodhand, 2009; Midgley et al., 2014: 9; Watt and Zepeda, 2012).

The measurement and definition of failed states is a contentious, political process that tends to reflect Northern values. It typically overlooks plural systems of governance that do not align with Northern ideals but which may, nonetheless, deliver degrees of security and stability to local populations (Kilcullen, 2013; Wilson, 2009). Inherently, such projects mobilize systems of representation that tend to render those states in the global South beyond the reach of Northern normativities. Notwithstanding this situation, however, it is possible to recognize examples where, among competing systems of governance, illicit, violent, organized criminal actors have assumed positions of political and economic prominence. Here, for example, Somalia is often cited as an example of a failed state where criminal sovereignties have underpinned extensive maritime piracy operations (Hastings, 2009; Hesse, 2011; Pham, 2011). However, to reduce the representation of Somalia to this single condition is insensitive to the multiple governances that have emerged during different periods under conditions of state stress there (Kilcullen, 2013: 66–69; Menkhaus, 2007).

More common than the complete or near total collapse of the state is the presence of governance "black holes" within otherwise functional states. These are areas at the subnational or local levels, referred to as areas or territories of limited statehood, spaces where the state is unable to maintain full authority (Krasner and Risse, 2014; Lilyblad, 2014; Risse, 2011). In such places "violent non-state actors ... often establish alternative governance structures, entailing ipso facto challenges to states' juridical territorial sovereignty" (Lilyblad, 2014: 74). Examples of such areas include the *favelas* of Rio de Janeiro (Lilyblad, 2014; McCann, 2006); the garrison communities, such as Tivoli Gardens, of Kingston, Jamaica (Kilcullen, 2013: 40, 63–66); and numerous borderland regions characterized by intense cross-border flows (Abraham and van Schendel, 2005; Costa and Schulmeister, 2005; Goodhand, 2009; Wilson, 2009). These territories are far from ungoverned. It is just that it is not usually states that are doing the governing there (Kilcullen, 2013: 96; Menkhaus, 2007). Such territories are common globally with up to two-thirds of all territorial-sovereign states incomplete in these ways (Risse, 2011: 2–9, cited in Lilyblad, 2014: 75; Wilson, 2009). As Krasner and Risse (2014: 545) argue, "the ideal-typical conception of a consolidated state is misleading. . . . This ideal-typical construct is far removed from the situation that exists in most of the world's polities."

The factors underpinning these plural and alternative governance contingencies are complex and diverse. Although areas of limited statehood are manifest at the local or subnational levels, and are often framed in predominantly endogenous terms, they are manifestations of multiple factors that often exceed single scales. Factors that are frequently cited as underpinning the emergence and persistence of areas of limited statehood include violence associated with competitive control of major transnational commodity and transport networks and local criminal markets (Daudelin, 2010; Kilcullen, 2013: 46); rapid, extensive informal urbanization processes where the state is largely absent (Davis, 2007; Kilcullen, 2013: 35-37; Muggah, 2014; Roy, 2011) or where the state unwillingly, willingly, or pragmatically cedes, negotiates, and/or competes for authority with other, nonstate, actors in these spaces (Brophy, 2008; Costa and Schulmeister, 2007; Cribb, 2009; Kilcullen, 2113: 94–96; Lee, 2008; Weinstein, 2008); inadequate or inept policing (Lilyblad, 2014: 78); limited governance capacity (Kilcullen, 2013: 48; Lilyblad, 2014: 81; Muggah, 2014); and the mobilization of popular cultural forms in opposition to state authority (Lilyblad, 2014: 82). Three points are worth making about these factors. First, in many instances, they exceed single scales. For example, while urbanization processes produce informality at the local scale, they are constitutive of "megatrends" operating and reshaping urban and rural life at the planetary scale (Davis, 2007; Gamba and Herold, 2009; Kilcullen, 2013). Second, factors operating at different scales may coarticulate. For example, local violent narcotics economies are sustained in ways that limit statehood by global prohibition regimes that locate their control beyond the reach of the state (except through forms of proxy governance negotiated with nonstate actors) (Andreas, 2005). Finally, areas of limited statehood do not seem to be explicable with reference to only singular factors. Rather, they seem to be characterized by the combination of a number of different factors that come together within territories. These points are significant in policy terms as they suggest solutions that respond to the multiple and multiscalar nature of the issue of limited statehood.

It is neither necessarily inevitable, though, that organized criminal economies will emerge in areas of limited statehood, nor is this absolutely a prerequisite for their formation. Such economies have been observed in strongly governed spaces, for example, where politicians embrace corruption, where criminal economic activities are able to evade state authority, or where organized criminal groups are able to embed themselves in state structures and processes (Levi, 2014; Weinstein, 2008). This has previously been noted in the case of China (Glenny, 2008: 366–367; Hudson, 2014: 786; Phillips, 2005; Scott-Clark and Levy, 2008), but examples also include the extensive production of cannabis in British Columbia, Canada (Glenny, 2008: 245–255), the development and embedding of mafias in the relatively civic region of Piedmont in Italy (Varese, 2011), and societies such as Japan where some criminal organizations enjoy a quasi-legitimate status and have socially legitimized roles in the mediation of business and civil disputes and the provision of goods and services (Glenny, 2008; Hill, 2005).

Combinations of factors, rather than singular causes, underpin the extensive development of organized criminal economies within regions. Single factors, even one as obviously important as limited statehood, appear to be insufficient to account for the emergence of organized criminal economies. In almost all cases, accounts of the development of organized crime within regions cite multiple factors in combination. These combinations, however, differ significantly between regions. For example, in his extensive empirical analysis of the migration of mafia groups beyond their home region, Varese (2011: 8) cites combinations of factors including the presence of booming economic sectors, inadequate state protection, demand for criminal protection, export-oriented economies, large illegal economies, and incentives for cartel formation as key to successful transplantation of mafias into new regions. These contrast, though, with those accounts of organized criminal economies that emphasize the economic marginality of their regional contexts within their accounts (Aning, 2007; Bagley, 2005; Castells, 2000; Glenny, 2008: 245–255; Hignett, 2005).

MULTISCALAR APPROACHES

Economic geographers have called for the analysis of economic processes to reflect the multiscalar ways in which these processes, such as those associated with globalization, unfold. Coe and colleagues (2007: 344), for example, argue that "any given set of practices is thus the outcome of a complex mingling of influences at multiple scales" and, with specific reference to criminal economies, Hudson has stated that "the contemporary illegal economy is grounded in a subtle interplay between activities at different scales" (2014: 780). However, as we have seen, it is rare to find examples of multiscalar approaches within the literatures of transnational organized crime. In being isolated from debates within economic geography, it would appear that the literatures of transnational organized crime have been somewhat immune to recent critiques that have questioned the appropriateness of single-scale approaches to understanding various dimensions of globalization. The limitations of much of these literatures is apparent when looking at, for example, Castells's widely cited accounts of the development of organized crime in Russia and Colombia across the 1980s and 1990s. His account of Russia locates its explanations largely at the national and international scales and within the realms of the structural transformations unfolding across Russian space following the collapse of the Soviet Union. His account points to the exploitation, by a variety of local, national, and international actors, of the institutional and regulatory failures that emerged during the first decade of Russia's chaotic transition to capitalism. Such macrostructural accounts have been the subject of a vigorous critique.

> The critique of Castells's approach, and the proliferating discourse about transnational organized crime, has pointed out that it is too general and lacks nuance to local conditions and historic contexts. . . . Castells's model thus overlooks the multiplicity of micro-practices and associations that, "while often illegal in a formal sense, are not driven by a structural logic of organization and unified purpose" (Abraham and van Schendel, 2005: 4). (Aas, 2007: 125)

In contrast, Svetlana Stephenson (2015) offers a reading of organized crime in Russia that is rooted among its local worlds and that advances a more ethnographically informed reading. Despite this alternative orientation, Stephenson's account is not blind to the national context of the street worlds she explores. It reveals, for example, a number of additional national factors to those present in Castells's account that lie beyond the economic and regulatory realms to which Castells looks. These included the senses of relative deprivation felt by the Russian working class that emerged both with the widening social differentiations of the 1970s and 1980s and especially following the collapse of the Soviet Union and the displacement that the latter caused to impact on them. These bred, particularly among young people, a willingness to turn to alternative social formations, particularly the gang. Stephenson also points to the loss of a collective sense of national identity in the post-Soviet period. Furthermore, within the economic and political spheres, the 1990s saw predation by both the state and corporate sectors, both of whom maintained a number of connections with organized criminal groups. Finally, the economies of the street became increasingly significant and were organized predominantly around small businesses that proved particularly vulnerable to exploitation and extortion by gangs (Stephenson, 2015: 41–46). Stephenson's account also makes room for the discussion of regional differences and particularities within the landscape of Russian organized crime, such as relative levels of economic development or deprivation, important in determining the extent to which Russian gangs were integrated into their local communities; the various economic legacies of the Soviet era in different regions; and local and regional histories of gangs and organized criminal activities (Stephenson, 2015: 46). Her account also speaks, though, of the relationships between scales, for example, recognizing how, as a consequence of the growing strength of the Russian state in the late 1990s and early 21st century, local criminal associations became more disorganized in form (Stephenson, 2015: 8). The most valuable contribution that Stephenson's account makes, however, much like that of Hobbs cited above, is through the wealth of local empirical detail that reveals the various "micro-practices and associations" (Abraham and van Schendel, 2005: 4, in Aas, 2007: 125) of organized crime that proved so elusive in Castells's account. Stephenson's account, then, reminds us of the value of incorporating multiple scales of analysis, the relationships between scales, and recognizing the combinations of factors that underpin the emergence of organized crime in different spatial contexts. Relatively, though, even with work of this sophistication, the orientation remains toward a single, rather than multiple, scale.

NETWORK ONTOLOGIES

One alternative to these scalar hegemonies is offered by the emergence of a set of approaches that have been labeled "network" ontologies (for discussions of the evolution of network ontologies of various kinds, key debates, and aspects of specific detail, see Bosco, 2006; Dicken et al., 2001; Ettlinger and Bosco, 2004; Murdoch, 1997, 1998; Sheppard, 2002). These approaches adopt more relational perspectives that focus on the connections that exist between "things," people, places, and objects; the way these come into being and are sustained; and the impacts they have on the places, people, and objects that are bound into networks. They offer attempts to transcend series of enduring dualisms that have long characterised debates within the social sciences. Such dualisms include such fundamental concepts as local–global, structure–agency, and nature–culture. The analysis that network approaches promote are sociospatial rather than locational. The unit of analysis is the flow or the connection rather than the more typical spatial units such as nation or region (Aas, 2007; Law and Urry, 2004).

The certainties of scale, that entities are locatable ontologically at a single scale, begin to dissolve when issues are approached in network terms. Network ontologies ask that we rethink the ways in which we view space. Places, rather than being considered bounded, discrete entities, are viewed as bundles of interconnections, nodes, within a variety of networks, thus rendering them both local and global simultaneously. This viewpoint has implications for the ways in which we understand phenomena that are manifest within specific locations. The regional case-study genre discussed above in the context of the literatures of organized crime has tended to privilege a particular unit of analysis and associated set of explanatory factors. The locational analysis upon which these accounts are typically based is characterized by the identification of endogenous and, on some occasions, exogenous, factors either within locations or their regional contexts. Regional context here is usually defined with reference to contiguous space. However, viewing places in networked terms raises the possibility that phenomena may be explained with recourse to connections to locations that may be distant in space rather than contiguously located. Network analysis, therefore, asks us to focus frequently on spaces that may be spatially distant but which are none-the-less closely bound together within networks of various types. It asks us to think of space not in Euclidean terms, but rather in networked or relational terms. This has been termed "positionality," whereby "conditions in a place do not depend on local initiative or embedded relationships across space . . . but on direct interactions with distant places" (Sheppard, 2002: 319). Places, then, may be spatially proximate but may have fewer meaningful connections between them than they have with more distant spaces.

By developing network perspectives we can begin to see how they provides a counterpoint to some of the structuralist accounts of transnational organized crime that give greater prominence to global economic contexts. For example, Sheppard's (2002) concept of positionality offers a rather different take on the conceptual mapping of global criminality that Passass (2001) explored in his account of the criminogenic consequences of global asymmetries. Passass argued that it was these asymmetries that generated interconnections between the distanced spaces of transnational organized crime. Positionality, however, argues the opposite, that it is the interconnections themselves that play a key role "in the emergence and persistence of inequalities within global economies" (Sheppard, 2002: 319). This maps out different theoretical and empirical avenues for transnational organized crime research than have existed to date. In ascribing agency to the interconnections between places, network ontologies argue that we should pay more attention to the interconnections that exist between places within networks. The networks that exist within the global economy of transnational organized crime illustrate positionality well. They form often significant, enduring connections between distant places that may be influential in shaping their developmental trajectories. The application of a specific network perspective, the global production networks (GPN) approach is discussed within Chapter 6.

The diversity of micro-opacities, for example, shipping containers, apartments, warehouses, laboratories, trucks, and vans that are exploited by criminal entrepreneurs and which were discussed above appear at first glance to be categorically local. However, to read these spaces through a local–global duality would be erroneous. Shipping containers, for example, are local and global simultaneously. They are local in the sense that their specific position, say at the bottom of a tower of other containers in the middle of a ship, makes them unlikely to be selected for inspection (Nordstrom, 2007: 160). At the same time, in sitting on an ocean-going cargo ship they are articulated by a global network of shipping lanes, their associated land-based distribution infrastructures, and global systems of registration, monitoring, and security. Similarly, the global mobilities of shipping containers and the commodities they move can only sustain illicit economies if there are "local," mobile, and immobile, spaces of opacity with which they are articulated, to receive trafficked goods once they are unloaded. We can argue also that these diverse microspaces of opacity are part of the constitution of globally significant economies, characterized by varying shades of informality, illicitness, and illegality (Samers, 2005; Williams, 2004, 2006) and that are not unconnected to contemporary global capitalism (Nordstrom, 2007: 15). As has been argued here and elsewhere (Hall, 2010a, 2010b, 2013; Hudson, 2014), it is vital to acknowledge these economies in critical accounts of contemporary capitalism.

Summary

There is not a universal cause of organized crime, a single factor that accounts for the development of criminal economies across diverse regional settings. This is not surprising given the empirical diversity characteristic of those activities to which the label "organized crime" has been attached. Nor, would it appear, is there a single factor within individual regions that can account for the development of organized crime there. Policies, then, that might imagine that there is this causal universality will inevitably be blunt instruments, failing to engage with at least some of the causes underlying the development of organized crime within places. Some factors, such as various forms of state contingency and regulatory asymmetries or opacities may be relatively more important, but they are not absolutely necessary in themselves and they can be achieved in multiple ways. There are layers of complexity here that policy seems to rarely recognize, let alone engage with. This, perhaps, offers a challenge to develop more plural policy models, both in terms of shaping policies toward specific contexts and in designing policy that articulates more with multiple factors underpinning the development of illegal economies within regions.

Conventionally, the development of organized criminal economies within places have been analyzed with recourse to hosts of factors conceptualized as local, national, or global in nature. However, this chapter has contended that to view these factors in this way underplays their multiscalar complexities significantly. Elsewhere it has highlighted that there is some debate about the ways in which characteristics in place and networked connections, of which places are part, relate together and their relative importance that remains unresolved within the literature. There appears to be much potential to apply the network perspectives cited in this chapter both in attempts to find new, innovative lenses with which to attempt to understand organized crime and also in the development of more effective policy responses. In doing so it is likely that the analytical edge of these approaches will be further refined.

The evidence suggests that, whatever the causes of the development of organized crime, they are simultaneously geographically specific, networked, and multiscalar. It is the geographically specific combinations of factors, the interplay between them and their positions within wider networks that bind spatially distance places together, that underpins this development. Thus, in approaching organized crime, it is important to recognize this geographical specificity. In policy terms, this suggests that universal policy approaches are likely to have differential outcomes across diverse criminal terrains and that policy models are likely to have only limited geographical transferability. The diversity of criminal terrains demands policies that are similarly diverse. The networked nature of criminal terrains further demands that policies do not solely act within places but rather are also oriented outward across networked space. Chapter 7 returns to questions of applying the perspectives outlined in this chapter in more detail.

CHAPTER 6

CRIMINAL MOBILITIES

INTRODUCTION

As the discussions in the preceding chapter showed, the markets associated with organized criminal activities are typically transnational, mobile, and networked. Spatialized readings of these markets tend to portray them as a series of turfs: spaces of production, manufacture, or origin; spaces of market exchange; or spaces of consumption. These spaces are articulated through connections along which the agents and commodities of these trades move or are moved. The attention paid to the various elements of these markets, however, is somewhat uneven. Prevailing accounts tend to focus predominantly on turfs, typically situating these analyses within readings of the factors, identifiable at a variety of scales, that account for the presence of criminal groups and their activities within these spaces (Hall, 2010a, 2012). Thus, discussions of the embeddedness of organized crime have tended to restrict themselves to either the ways in which organized crime groups emerge out of the socioeconomic contexts of particular, often turbulent, regions and, despite extending their geographical reach over time in some cases (Varese, 2011), retain strong connections to their home turfs; or the ways in which relatively lucrative criminal markets can suffuse the economic, political, and institutional spheres of their regions (Castells, 2000: 183; Grascia, 2004;

Hill, 2005; Kleemans and van de Bunt, 1999; Madsen, 2009; Scalia, 2010; van de Bunt and Siegal, 2003: 4).

This is not to say, however, that analysis of the spaces of movement, transit, and distribution associated with these markets are entirely absent. There has been, for example, considerable academic, practitioner, and journalistic attention focused on the transit of illicit and illegal commodities where they have affected violent outcomes and/or significant economic or geopolitical transformations, for example, in the cases of the Balkans, Mexico, and West Africa, key transit zones in the smuggling of cigarettes and the trafficking of cocaine and heroin (Aning, 2007; Brophy, 2008; Ellis, 2009; Glenny, 2008; Saviano, 2016; Vulliamy, 2010; Watt and Zepeda, 2012). In contrast, other authors have spoken of the ways in which political and development outcomes have emerged along the routes of criminal commodity movements, spaces where often the reach of the state and development agencies is tenuous or incomplete (Bhattacharya, 2005: 118; Goodhand, 2009: 22; Nordstrom, 2007; van Schendel, 2005: 58). There is also an important body of work within the social network, gravity, and latent space models approaches, which is summarized later in this chapter, which has sought to reveal the broad determinants of the spatialities of networks of illicit trade and transit (see Berlusconi et al., 2017), if not so much the effects these networks have on the spaces through which they pass. From a very different methodological tradition, we might also recognize the obvious potentials of alternative anthropological and ethnographic "follow-the-thing" approaches (Cook, 2004; Cook and Harrison, 2007), although they have yet to be applied to the study of illicit or illegal objects. However, accounts of spaces of transit and distribution located along criminal networks, and the relationships between these networked connections and the spaces through which they pass, is somewhat patchy and incomplete (Berlusconi et al., 2017: 2) and is dispersed across a variety of literatures and tend, overall, to be subservient to the more territorialized readings of spaces of production and consumption within the literatures of organized crime. Drawing together insights from literatures that have foregrounded analysis of the spaces of flows, as this chapter seeks to do, suggests, despite the emergence and consolidation of a number of networked, mobile, and postregional ontologies in recent years (Aas, 2007; Law and Urry, 2004; Nordstrom, 2007; van Schendel, 2005), that understandings of the effects of the connections *between* the most prominent spaces of major criminal markets remains somewhat underdeveloped and undersynthesized.

The prevailing grounding of accounts of criminal markets within their regional geographies of production, exchange, and consumption, rather than equally within the connections between these spaces, has tended to advance representations that run the risk of imbuing a sense of ontological separation between criminal commodity mobilities and the spaces through which they pass. It is, a little, as if all that really matters occurs within bounded regional spaces and the connections between them are little more than forms of conveyor belt, all be they ones, as it is occasionally acknowledged, that are shaped by various geopolitical and geographical contingencies. For example, approaches such as commodity chain analysis speak of the organization of economic flows and their "local development outcomes in those areas where the chain touches down" (Bair and Gereffi, 2001: 188, in Mackinnon and Cumbers, 2007: 150). However, it is argued here that a more useful starting point is to consider such connections and their associated mobilities as entities that are grounded (they touch down) and are active at all points. As van Schendel (2005: 46) argues, "These flows do not move in thin air and they are not disembodied; we need to incorporate the social relations of transport and distribution, and their spatiality, in analysis of global rescaling." Thus far, however, there has been precious little analysis that has done this and sought to reveal the grounded realities of criminal commodity mobilities, or indeed commodity mobilities of any kind (though for innovative exceptions, see Birtchnell, Savitzky, and Urry, 2015; Cook, 2004; Cook and Harrison, 2007), across their spaces or the economic, social, and political reciprocities they are entangled within and across the ranges of territories with which they articulate (Kilcullen, 2013: 112–113).

This chapter aims to speak to this underdeveloped aspect of the literature through the case of criminal commodity movements, connections that articulate the regional geographies of criminal markets. In doing so, though, it aims to speak also to the wider literatures associated with various relational ontologies within economic geography and the social sciences more generally. It will do this primarily by excavating and drawing together a variety of literatures that reveal aspects of the transformative relationships between criminal commodity mobilities and their spaces and from these aim to build a more nuanced understanding of the connections that bind together criminal markets. The chapter now moves to a broad discussion of networked criminal mobilities and the determinants of their geographies before moving on to consider the representations of illicit and illegal commodity mobilities and their effects.

NETWORKED CRIMINAL MOBILITIES

As the previous chapter argued, as organized criminal groups engage with the transnational markets associated with the transit and sale of commodities such as narcotics, trafficked people, and counterfeit goods, they have inevitably come to demonstrate international orientations of various kinds (Aas, 2007: 5). However, this does not confirm the specter of transnational mafias alluded to in Chapter 1 (Levi, 2002: 907). Rather, the transnational nature of many of the trades now central to major criminal markets have caused groups to seek collaborative, networked arrangements with other criminal groups across international terrains (Galeotti, 2005a; Hobbs, 2001; Hornsby and Hobbs, 2007; Kenney, 2007; Wright, 2006). This appears to have emerged as an almost standard organizational arrangement for groups operating within these markets. As Berlusconi and colleagues (2017: 10) argue with regard to one transnational illegal market:

> Heroin traffickers tend to travel relatively short distances when moving drugs from country to country. . . . This can increase the time needed to move the load to its destination but, more importantly, keeps the level of sophistication required for the shipments simple, in turn, reducing the risk of interception and likelihood of arrest for traffickers.

It is important, though, not to attribute the transnational networks associated with illicit and illegal trades with a kind of weightlessness. Reuter (2014: 34), for example, talks of the imperfect environmental and market knowledges of drug traffickers with regards to opportunities and costs associated with potential trafficking routes, while Goodhand (2009), with specific reference to the Afghanistan opium industry, highlights the embeddedness of illicit networks in their societies, both of which impede rapidly responsive mobilities. The development, maintenance, and responsiveness of illicit networks requires the investment of considerable criminal labor and they are far from universally or automatically successful (Hobbs,

2013; Varese, 2011), something that could be more widely explored empirically and acknowledged within the literature.

Practically, networks offer more achievable ways of generating the international orientations that criminal groups require within transnational economies than more centralized transnational structures. Varese (2011) has demonstrated the difficulties that criminal groups face in establishing and maintaining any successful presence in an overseas territory, something that would be a prerequisite for developing a centralized, transnational structure. Given these difficulties, it is hard to imagine a criminal group sophisticated enough and with the resources to maintain a globe-spanning presence. Furthermore, the coordination of the commodity chain, and the necessity to negotiate the authority and scrutiny of both state and nonstate actors at all points, would seem to be a challenge beyond the capacities of the small- and medium-sized criminal organizations that are characteristic of illicit economies internationally without these networked arrangements. Seaports, for example, are key spaces in the global mobilities of illicit commodities. Notwithstanding the presence of state security agencies within these spaces, they are also spaces that are rich with agents of nonstate authority (Brooks, 2012; Glenny, 2008; Madsen, 2009; Nordstrom, 2007; Saviano, 2008). It is hard to imagine competitive incursions from a single criminal organization across multiple, widely dispersed international port spaces all being successful and remaining undetected or unanswered by either state, international, or illicit authorities. Rather, criminal groups pragmatically have opted to collaborate with extant, embedded illicit authorities to mobilize their economies.

The criminological and sociological literatures of contemporary organized crime record numerous examples of this form of networked arrangement operating across multiple criminal markets. These include the extensive economies of transnational cocaine trafficking (Saviano, 2016); the production of cannabis in the United Kingdom and the Netherlands (Schoenmakers et al., 2013; Silverstone and Savage, 2010); the transit of coltan, a key component of mobile phones and other electronic devices, which is sourced to a significant degree through opaque mechanisms from the former conflict zones of the Democratic Republic of the Congo (Glenny, 2008); the cocaine-producing and trafficking *cartelitos* of Colombia (Bagley, 2005; Glenny, 2008; Kenney, 2007; Wright, 2006); networks of organized retail theft that have been identified as a source of terrorist financing (Madsen, 2009); the geographically distributed, cooperative structures observed within cybercrime economies (Glenny, 2008; 2011; Madsen, 2009), although the limitations of data relating to this economy should be acknowledged (Lusthaus, 2013); cases of routine collaboration between organized crime groups in the United Kingdom and groups in other countries (Kirby and Penna, 2011: 186); and cooperation between indigenous Italian organized criminal groups and those arrived from China, and more recently Nigeria, in the economies of counterfeit goods production and smuggling, waste disposal, human trafficking, and money laundering (Madsen, 2009: 51–53). Here, despite the geographical, cultural, and linguistic differences between these groups, they have found ways to operate in mutually beneficial, rather than competitive, relationships. For example, the presence of Italian organized crime groups in the country's ports has been deployed to facilitate the entry of smuggled and trafficked goods from China (Madsen, 2009: 52). This is captured in the journalist Roberto Saviano's discussion of the port of Naples in Italy and its role, and that of its dominant criminal organization, the Camorra, as a key node within networks of the global trade in counterfeit goods:

> The Far East, as reporters like to call it. Far. Extremely far. Practically unimaginable. Closing my eyes, I see kimonos, Marco Polo's beard. Bruce Lee kicking in mid air. But in fact this East is more closely linked to the post of Naples than to any other place. There is nothing far about the East here. It should be called the extremely near East, the least East. Everything made in China is poured out here. Like a bucket of water dumped into a hole in the sand. (Saviano, 2008: 4)

Chapter 2, with reference to transnational illicit markets in illegal drugs, discussed explanations for the distribution of sites of the production of drugs in relation to their markets, but did not say anything about the ways in which the routes they take to market are determined and shaped. Indeed, it has been recognized that "there remains little systematic research on drug flows between countries" (Berlusconi et al., 2017: 2). Although drug routes worldwide are generally well known, in their broad outlines at least, their determinants have been, perhaps, more assumed than rigorously tested. Berlusconi and colleagues (2017: 1–5), though, do provide a useful summary of a small body of research utilizing social network, gravity, and latent space approaches, as well as those from the more general criminological literature, that have sought to uncover the determinants of the geographies of illicit commodity networks. The insights from these literatures include the recognition of significant differences between legal and illegal trade networks. Boivin's research (2013, 2014a, 2014b), for example, shows that illicit drug trafficking networks tend to be more structurally ad hoc than legal trading networks, while core economies within the legal economic sphere are more peripheral to transnational illicit drug economies. Furthermore, the overlap between different illicit trafficking networks may be limited, as Chandra and Joba (2015) demonstrate in their analysis of international heroin and cocaine networks. Other research highlights the influence of factors such as the economic "mass" of countries both in terms of their licit and illicit economies; regulatory, legislative, and governmental asymmetries; distances between potential trading countries in illicit networks measured in both geographical and social terms; and the risks and potential profits facing criminal entrepreneurs. This research has spanned illegal and illicit markets including the underground flows between U.S. states (Wiseman and Walker, 2017) in activities such as drug smuggling (Reuter, 2014), money laundering (Walker and Unger, 2009), human trafficking (Akee, Basu, Bedi, and Chau, 2014) and transporting illegal guns (Kahane, 2013). Berlusconi and colleagues' research is innovative methodologically but also conceptually in that they allow their hypotheses to be informed by the insights of criminological literature from outside their own methodological compass. Here they provide a model of interdisciplinary dialogue that would repay more general deployment. Their own analysis confirms the significance of social and geographical proximity and levels of corruption in shaping illicit trading networks (2017: 8–11).

The routes that illicit commodities take to market, then, are diverse and complex, and they emerge out of the interplay of the qualities of the commodities themselves, their materialities, and the often fluctuating degrees of licit and illicit, legal and illegalness associated with these mobile objects, for example, and various conditions across space. As previous chapters have demonstrated, the scale at which analysis is conducted is also crucial in shaping our knowledge of the geographies of illicit mobility. For example, Chandra, Yu, and Bihani provide a rare example of research into subnational drug movements in their analysis of ecstasy (MDMA) trafficking patterns in the United States, where they identify the significance of border and coastal cities within this market (2016, in Berlusconi et al., 2017: 2), something not apparent in analysis at the national scale and beyond.

CRIMINAL COMMODITY MOVEMENTS AND THEIR SPACES

Clearly, then, illegal flows are not external forces that, arrow-like, fly past supine borderland societies. On the contrary, they are actively domesticated and incorporated into borderland projects of scalar structuration.

—VAN SCHENDEL (2005: 55)

Figure 2.1 displayed the major global trafficking flows of opiates. It showed that a significant proportion of these move out of Afghanistan, where the greater portion of the world's opium is produced, along the Silk Road route into southern Europe. Of course, this map says nothing about the embeddedness of these commodity movements within the spaces and societies through which they pass. Such insights are the preserve of more ethnographic accounts. Bhattacharyya (2005: 117–118), for example, discusses these grounded realities of the illicit globalizations of the drugs trade, highlighting the ways in which these illegal commodity movements utilize very traditional modes of transport. In this case, the movement of these commodities clearly becomes embedded in the lives of those societies through which they pass: "This is the manner of a global integration that touches a range of ordinary lives across nontechnological and barely developed localities and provides employment for some of those who have little access to other kinds of work in the global economy" (Bhattacharyya, 2005: 117).

Simplistic renditions of such transnational drug flows tend to equate them with lawlessness, insecurity, violence, and instability. However, logically and empirically, there is much to challenge such assumptions. Universal prohibition has made outlawed narcotics highly lucrative commodities. Presumably, if they have any rational decision-making capacity whatsoever, the very conditions that those involved in the transit of narcotics would seek to avoid are those of lawlessness, insecurity, violence, and instability. As Goodhand (2009: 20) argues, "Violence is bad for business and can be understood as a sign of market dysfunction," and further that "this case study [of the Sheghnan region of Afghanistan] calls into question an influential policy narrative that associates lootable resources such as drugs with insurgency, warlordism and state collapse" (Goodhand, 2009: 23). Indeed, where such conditions do tend to prevail in drug transit zones, they are typically the result of external intervention disrupting, admittedly fragile but often effective, equilibriums between competing actors (Watt and Zepeda, 2012). Rather, what was hinted at in Bhattacharrya's quote above, but which has been generally little discussed, perhaps because it is an unpalatable truth for development agencies and transnational institutions, is the potential for the movements of drugs, and indeed other illicit commodities, to deliver conditions of stability, security, order, and development to otherwise hard-to-reach regions. While the impulses of those involved in the drug trade might be a mixture of the predatory, the instrumental, and perhaps even the altruistic, what is unarguable is that stable conditions favor the transit of lucrative commodities (licit or illicit) much more than conditions of instability. Goodhand's (2009, 2011, 2012; see also Goodhand and Mansfield, 2013; Goodhand and Sedra, 2015) analysis of the Afghanistan opium trade does offer a valuable corrective to the more violent renditions of the effects of transnational drug movements.

Nordstrom's ethnographic, mobile explorations of international extralegal networks echoes those of Goodhand. She outlines the development outcomes apparent within a host of illicit trades and their associated networks:

> And while they [extralegal businesses] clearly benefit personally, their actions often bring development to their communities. In areas where infrastructure is weak and governments can provide little in the way of social services, it is people like them who help rebuild regional industry, donate to health and education, and bring critical resources for the citizenry. This is the ultimate contradiction defining the extra-legal: it can act as both harmful profiteering and positive development—sometimes simultaneously. (Nordstrom, 2007: 99–100)

What Goodhand above outlines is not an industry based on violent, predatory banditry, although that certainly does exist within it, but rather one that is integrated into the social, economic, and political dynamics of Afghanistan's border regions. The organization of drug smuggling within the regions he discusses has evolved through time from a cottage industry characterized by a large number of small couriers in the period prior to 2001, to something more professional and vertically integrated during the significant expansion of the Afghan drug economy after 2001. This professionalization of the drug economy there, and the consequent changes in its modes of organization, primarily reflect international trends in the organization of drug networks at the time. Official, hegemonic representations of illicit and illegal commodity movements, some of which are discussed in the next section, are based largely around commodity volumes and directions of movement. Consequently, they can say little about questions of the organization of these flows and their integration into the spaces through which they pass, and can only provide somewhat denuded and ahistorical representations of these shifting mobilities.

Goodhand's work has shown, though, that these drug transit networks were not simply reflective of their various contexts, but were also active in shaping them in a number of ways. These included a range of economic benefits to the regions through which these drugs were trafficked, including visible infrastructure developments; increasing investment in licit businesses as drug monies were recycled through regional economies; concomitant increases in rural wages; increased availability of credit for poor households; and growing demand for consumer products (2009: 22). However, Goodhand charts how the effects of the movement of these commodities through the borderlands of Afghanistan have exceeded the solely economic and included the transformation of socioeconomic relations, including the emergence of a capitalist class associated with the drug trade, and the transformation of a host of other intergroup relations. It has also affected geopolitical relations between regions in Afghanistan and political relations across the nation. Goodhand's description of these trades is of economic activities and mobilities that are deeply embedded in the regions that they touch. It helps to give lie to the myths of disembodied flow evoked in their cartographic representation in numerous official accounts. As Goodhand (2009: 21) argues,

> The drugs industry then is far from anarchic. Rather, it is underpinned by various forms of state and non-state regulation . . . including indebtedness, addiction, bribes, gifts, violence, marriage and public displays of benevolence. It involves collaboration and collusion between state and non-state, which can become increasingly formalised and even institutionalised in the post-Bonn period, to the extent that today, the removal of key actors within the system would, arguably, have little effect on the overall functioning of the system—which makes nonsense of arguments that the drugs industry can be countered by removing a few "bad apples."

Illicit and illegal movements, then, are "actively domesticated" (van Schendel, 2005: 55) in these and many other ways within a variety of spaces and societies through which they move. To return to the point asserted at the start of this chapter, they would profitably be thought of as grounded at all points. One approach that provides some conceptual purchase in attempts to understand networks of commodity movement within their spatial contexts is the global production networks (GPN) approach (Coe et al., 2008; Hudson, 2005). This is an approach that developed out of a number of earlier approaches such as global commodity chains and global value chains, both of which, like the GPN approach, sought to understand "the nexus of interconnected functions, operations and transactions through which a specific product or service is produced, distributed and consumed" (Coe et al., 2008: 272). The GPN approach builds on these earlier, "more restricted" (Coe et al., 2008: 272) approaches in a number of ways. Specifically, while like these earlier approaches and indeed all relational approaches, it makes network connections its primary unit of analysis, GPN highlights the spatial and temporal complexity of network connections. Furthermore, it is an approach that examines networks not as disembodied entities, but rather as movements grounded within a variety of spatial contexts. GPN approaches offer clues to a number of the issues that this chapter explored and have been utilized to explore the networks that move illicit commodities between countries, most notably in Brooks's work on the secondhand car (2012) and fashion trades (2015).

The former provides an ethnographic reading, rooted in a GPN approach, of the roles of corrupt practices within the used-car trade from Japan to Mozambique and the ways in which these are obfuscated through neoliberal discourses of "development success" (Brooks, 2012: 80). Here are revealed the complex, grounded, and multiple networks that belie the apparently smooth contours and unimpeded flows typical of official diagrammatic representations of transnational commodity networks. The movement of used cars from Japan to Mozambique, as is typically the case with transnational flows of many kinds, are the product of opportunities opened up by asymmetries, in this case, between the regulatory regimes of the two countries. This, however, only offers a starting point in understanding these commodity mobilities, their dynamics, and their impacts. Japanese used cars are not simply loaded onto ships, sailed to a convenient major port (Durban, South Africa), and then driven, unimpeded, to market. Instead, GPN approaches are sensitive to the fragmented, contingent nature of supply chains and the ways in which they are embedded in "disparate place-based social, political, legal and economic conditions in the terrain of global capitalism" (Lane and Probert, 2006, in Brooks, 2012: 81). Although the broad contours of this trade are outlined by international regulatory asymmetries, its specific routes and characteristics are reflective of a number of historical and local contingencies. For example, well-resourced Pakistani trading families play key roles in the import of used cars through South Africa. This is a development of their roles in earlier used-car importation networks to Pakistan, which emerged at a time when the Pakistani government insisted that used cars imported into the country had to have been owned by Pakistanis based overseas. This trade was subsequently banned by the Pakistani government, forcing these entrepreneurs to deploy their expertise across other used-car trade networks. Thus, the network is embedded within and shaped by a particular constellation of specific economic, regulatory, social, ethnic, linguistic, and gendered relations including Pakistanis, South Africans, and Mozambicans in Durban. They are also embedded in the fragmented local geographies of the Durban port-scape:

> The Pakistani used car lots, where only foreign passport holders can enter, are part of the duty free "international" port landscape and are separated from South African economic space. Transport hubs, such as these are key nodes in the geography of global capitalism and mediate and re-configure commodity flows (Sidaway, 2007). There are many Japanese used cars passing through this node. Cars are jammed into the lots, parked bumper to bumper. . . . They arrive from Japan tightly packed in shipping containers. Unloading is a rough process and cars are occasionally bumped and dented. Single used car units are relatively high-value commodities but their treatment in the lots, inside enclaved port sites where space is restricted, is indicative of how individual cars are part of a very large-scale and capital-intensive process. (Brooks, 2012: 84)

However, there is also reciprocity at play here between these commodity movements and their spatial contexts. While the relations outlined above shape this particular transnational commodity movement, they are also, in turn, shaped by it. Most notably, it is this particular movement of commodities through a specifically regulated national space that establishes and sustains sets of power asymmetries between the different groups who come together at points along this network. For example, the onward movement of used cars from Durban, through Swaziland, to their final market in Mozambique is important here. Mozambicans have been stopped from transporting the cars themselves by the South African local authorities. This transport has now become something in the gift of Pakistani used-car sellers and their South African contacts (Brooks, 2012: 86). The corruption evident at points along the network, deployed to elude the official scrutiny of state institutions and to negotiate lower tariffs by used-car traders, is evidence of these reciprocities between this commodity flow and its context. While corruption facilitates this trade, the trade, in turn, ensures that shifting power asymmetries and socioeconomic relations emerge and corruption becomes more deeply embedded within the cultures and practices of the institutions and actors involved. Thus, we have a commodity movement that is conditioned by international regulatory asymmetries and specific historical and geographical contingencies. It is socially embedded across multiple sites and both conditioned by and conditioning of these social relations.

* * *

Carolyn Nordstrom, in *Global Outlaws: Crime, Money and Power in the Contemporary World* (2007), provides an anthropological perspective on the global economy informed by an ethnographic method. She literally walks, drives, and sails the pathways of transnational illicit and illegal trade around the globe. The result is a peopled account of the contemporary global economy that stands in contrast to the more abstract accounts that have tended to prevail within disciplines such as economics, economic geography, and political science, where the ethnographic approach has attained less methodological purchase. Among the many insights into the centrality of these extralegal trades, in their many forms, to the contemporary economic and political mainstream that Nordstrom develops from her empirical sources is the idea of trade routes as markets. This crops up at a number of points within the text and is articulated in places by some of the respondents in the book, many of whom are extralegal traders or those charged with monitoring or addressing the issue of cross-border extralegal trade. While empirically evident throughout the text, there remains room for further conceptual development of this idea.

The idea of trade routes, of commodity movements, as markets is not one easily captured within prevailing theoretical accounts or, indeed, cartographic representations. Approaches such as global commodity chains and global value chains are hindered both by a tendency to underplay the complexities and heterogeneities of commodity movements and to see them only as intermittently grounded in the spaces through which they pass. There is, though, perhaps room within GPN approaches to accommodate the idea. As Brooks's (2012) account of the Japan–Mozambique secondhand car trade shows, GPN approaches are more attentive to the grounded realities and the complexities of commodity movements, particularly where they are located, wholly or partly, within the realms of the extralegal. There is clearly room to expand the deployment of GPN approaches across other aspects of extralegal trade and to focus more explicitly on the idea of these routes as markets, as revealed in Nordstrom's work and further, given that extralegal trades are significant, yet underappreciated aspects of the contemporary global economy (Nordstrom, 2007: 164, 2010: 173), to compare, more systematically, the spatialities and organization of commodity movements within the legal and extralegal realms, while noting the limitations of such discrete categorization of these movements.

Illicit and illegal commodity movements are characterized by the diversity of commodities that move together along them. It is rare to find uniform, undisturbed extralegal commodity movements. As well as the diversity of illegal and illicit goods moved within shipping containers that she describes, elsewhere Nordstrom records the movement of drugs, cigarettes, beer, diamonds, minerals, and electronic consumer goods within consignments of poached fish (as in illegally obtained, rather than gently cooked in liquid) such as abalone, crayfish, and Patagonian toothfish (2007: 8, 106–107). Often the sale and exchange of these goods does not take place with a single trader in one location but rather with a host of other traders at various points along trade routes. As one of her respondents, a senior official at the World Customs Organization in Brussels, Belgium, reveals, "Ships can stop anywhere at sea, meet up with any other ship at sea, exchange just about anything with anyone at sea—and who is there to monitor? They're floating supermarkets. This is huge—and I can't underscore enough the word huge—business" (in Nordstrom, 2007: 107). Criminal commodities movements, then, seem to fragment at multiple points of exchange within the networks along which they move, to move through a host of trading spaces, and to be exchanged multiple times between origin and market. The concept of commodity movements as markets, then, that emerges through Nordstrom's analysis reveals them to be complex, fragmenting, highly peopled, and grounded entities. These more critical views of such commodity mobilities suggest alternative policy responses to those that prevail currently. As this book has concluded at a number of points, once again they suggest, perhaps, policies that are turned inward toward the operations of the economic mainstream, rather than outward toward somewhat mythological terrains seemingly located beyond it.

The sense of ontological disembodyiedness that has tended to pervade accounts of criminal commodity mobilities is not entirely due to their being somewhat overlooked in empirical accounts of criminal markets. It is also, in part, a consequence of their prevailing representation within official accounts, representations that are worthy of further critical attention. It is to a discussion of the nature of representations and their effects that this chapter now turns.

REPRESENTATIONS OF ILLICIT COMMODITY MOVEMENTS

Maps filled with conspicuous arrows claiming to be scaled representations of illegal flows have been used to great effect to propagate particular ways of understanding spatial movements that lack state authorization.... When it comes to understanding illegal flows, their bold arrows hide more than they reveal. Usually they are quick stopgaps, hiding our lack of detailed knowledge, dramatizing and simplifying processes that we understand at best in outline, and forcefully pushing interpretations that need more careful consideration.
—VAN SCHENDEL (2005: 42)

A key phrase in the quote above is "simplifying processes that we understand at best in outline." There has been a taken-for-grantedness about the mobilities associated with the criminal economy within the official accounts that van Schendel refers to. Understanding them more than in outline reveals effects that are not captured in their prevailing cartographic representations and the discourses that emerge from these. The tendency to fail to fully acknowledge the groundedness of criminal commodity mobilities within the spaces through which they move, an aspect of the "understanding only in outline" above, is reflected in cartographic discourses of these illicit mobilities that permeate, particularly, accounts of organized criminal markets that emerge from state agency and transnational institutional sources (van Schendel, 2005: 40–41). Maps of illicit movements are often a key component in the official representation of transnational criminal markets (see, e.g., the United Nations Office on Drugs and Crime's annual *World Drug Report*). Formally, such maps are typically defined by official state borders that enclose the white spaces of nations. Across these spaces are lain a series of arrows of varying thickness and shade that represent the movement of illicit commodities across space. However, there are extensive and long-standing critical literatures of cartography that emerged first within cultural geography and subsequently within geopolitics (Crampton, 2001; Crampton and Krygier, 2005; Harley, 1988; Moore and Perdue, 2014; Pickles, 2004; Wood et al., 2010) that remind us that such maps are not innocently mimetic of an external reality, but rather stand as texts, inscribed with power and meaning that actively constitutes this reality (Crang, 1999: 60). Such representations, then, encode a visual discourse of illicit and illegal markets that captures nothing of the grounded realities of their connections. Given the prominence of the texts within which these cartographic tactics are deployed, such failures affect widely held perceptions of these movements and, following the arguments laid out in much critical cartography literature, help buttress official agendas.

Thus, these official renditions of the movement of illicit and illegal commodities around the world offer discursive understandings of these processes that are mobilized through a series of widely deployed visual motifs and textual metaphors. These elements typically work in mutually constitutive ways to reinforce prevailing understandings of the mobilities associated with transnational criminal markets. The official documents out of which these discourses emerge are premised on an intention to make visible what they purport to be invisible or hidden from view. For example, the following quotation is taken from a United Nations Office on Drugs and Crime threat assessment looking at the transit of opiates through northern Afghanistan and central Asia:

> Despite improvements to customs controls and the large-scale coverage of border guards, the majority of Northern route opiates continues to flow nearly uninterrupted into Tajikistan. Both large, well-organized groups and small entrepreneurs appear to be engaged in trafficking. Entrenched corruption and the strength of criminal organizations in Tajikistan make this flow largely *invisible* relative to its importance.
> The recent arrests on narcotics smuggling charges of high-ranking Tajik officials within the department for combating drug trafficking may be one indication of this *invisible* traffic. (2012: 13; emphasis added)

Thus, a language of invisibility and concealment tends to permeate accounts of the movement of illicit and illegal commodities within official documents. Consequently, these documents seek to challenge the apparent invisibility of the trades of which they speak, typically through references to the invisibility of criminal commodity movements being juxtaposed with maps that purport to reveal these commodity networks as arrows indicating movement across oceans and territories (see, e.g., Figure 2.1). However, both the intention (to make visible what is apparently invisible) and the discursive tactics through which this intention is realized are open to critical scrutiny. One who has begun this endeavor is van Schendel (2005), who has evolved a critique of the disembodied cartographies of illicit flows that emerges from his more grounded analysis of borderlands, their societies, and organization and the ontological questions that borderland perspectives open up.

A fundamental premise of official discourses of criminal commodity movements, then, is that they are invisible in an ontological sense, that they move unseen. This sense of invisibility, however, is constructed through an official, state-centric gaze. These movements are invisible in that they evade the gaze of the state and transnational institutions and do not appear in official statistics and accounts of trade. Where such movements are recorded officially, they typically appear in accounts of interceptions of criminal goods at various points in transit (see, e.g., United Nations Office on Drugs and Crime (2010b: 178), which are inevitably only partial glimpses of more extensive movements. Such mechanisms of surveillance, interdiction, and reporting discursively construct a world of elusive, shadowy movements that only occasionally come into the light of state scrutiny and law enforcement. These movements are not fully invisible to the state, then, but official knowledges of them are inevitably incomplete. However, this is not the same as ontological invisibility. For example, these mobilities do not appear to be fundamentally more invisible, in an ontological sense, than a host of others typical of the contemporary global economy that move between corporations, states, and transnational institutions. These movements include the movement of licit commodities within global networks of trade, roughly one-fifth of which, as noted in Chapter 5, routinely move undeclared (Nordstrom, 2007: 172); the ambiguous geographies of waste and e-waste recycling (Grant and Oteng-Ababio, 2012; Lepawsky and McNabb, 2010; Urry, 2014); global financial flows of both a licit and illicit stripe (Lewis, 2014; Sharman, 2011); the now global movements and trades in secondhand goods such as cars and clothes (Brooks, 2012, 2015) and people, for example, in the form of political prisoners who have been the subjects of secret rendition flights (Paglen and Thompson,

2006), and illegal migrants who service the economies of the global North (Aas, 2007: 40). The deployment of the metaphor of invisibility to describe the commodity flows associated with criminal markets should be seen within the context of wider discursive fields in which a host of other movements, despite being empirically comparable, are not necessarily so labeled. Thus, the discourse of invisible criminal mobilities that emerges out of these texts inscribes uneven ontologies across contemporary transnational movements that are associated with multiple agents, corporations, organizations, and institutions. This is a discursive construction of the licit and illicit that is reflective of current policy views that are unevenly directed toward tackling these mobilities. Nordstrom (2007: 164–165) argues for an alternative response: "The answer, it would appear, is in making visible the entire system of un/regulated and il/licit activity. The multinationals as well as the cartels; the respected corporations as well as the street criminals; the governments breaking sanctions for arms and supplies as well as the organized crime rings."

Furthermore, it is worth remembering that these mobilities are not actually invisible. Illicit and illegal commodities do not somehow move unobserved by human eyes, as if they were untouched by human hands. They are grounded in various infrastructures of contemporary commodity mobility (shipping containers, trucks, cars, camels, packing cases, rucksacks, airplanes, people, etc.). Illicit and illegal commodities might be moved within the formal infrastructures of contemporary commodity mobility and travel, for example, by being hidden among licit commodities within shipping containers (Martin, 2016; Nordstrom, 2007: 8; Urry, 2014), or being transported concealed on or within the bodies of drug mules on airplane flights (Fleetwood, 2014). These infrastructures are assemblages of materials, technologies, information, texts, surveillance, regulation, and people rather than hermetically sealed tubes through which commodities glide. Illicit and illegal commodities are directly and indirectly observable at many points within these assemblages and connect with the realms of the state and civil society and licit economic circuits at many points and in many ways along their transit (Nordstrom, 2007).

The complex mobilities associated with criminal markets are further rendered, within official accounts, through the deployment of a metaphor of *flow*. Van Schendel (2005) has been very critical of the deployment of such metaphors to represent such criminal mobilities. These, he argues, are deployed in ways that evoke images of alien threat through their contrast with national borders, whose representation in official texts, emphasizes notions of fixity and vulnerability. "In the discourse on illegal flows, agency rests with the flows. They are described as permeating borders, subverting border controls, penetrating state territories, seeking markets and finding customers. Borders, on the other hand, are presented as passive, vulnerable and reactive" (van Schendel, 2005: 41).

Metaphors of flow are also particularly amenable to generating watery rhetorics of flooding, drowning, and of barriers being overwhelmed (van Schendel, 2005: 39–40) by, in this case, movements of illicit and illegal goods and people. There is, however, nothing natural about the discursive associations that have arisen between criminal mobilities and the idea that they constitute flows which, in turn, may generate floods. These are metaphors, though, that have been deployed for some decades at least. Knepper, for example, quotes the British Conservative Member of Parliament Mary Berwick Ward, who attended the Seventh Assembly of the League of Nations in 1926. Referring to the deaths of over 1,700 drugs addicts in the city of Harbin, Manchuria, she argued, "Experience had shown that when any one point of the globe became a centre for illicit traffic ... all countries were in danger of being flooded by illicit drugs from the region in question" (in Knepper, 2011: 131). At this time, however, such associations appeared to be part of evolving and contested, rather than hegemonic, framings of illicit and criminal mobilities (Hall, 2017).

To be at all effective, such metaphors must possess some intuitive appeal. However, they should not be treated naively. The use of metaphor as a means of legitimizing political positions has been long exposed through scholarship in the critical geopolitics tradition (Hepple, 1992; Yanik, 2009). The deployment of the metaphor of flow here, often used in association with the cartographic equivalent the arrow, can be situated in that long tradition of states locating organized criminal activity as an external threat poised to overwhelm a pristine national territory, rather than it being viewed as, at the very least in part, an endogenous, locally grounded issue (Aas, 2007; Varese, 2011: 27; Woodiwiss, 2003; Woodiwiss and Hobbs, 2009). In this instance, the use of the metaphor of flow to represent these movements has gained traction within official literatures, which is in part due to its intuitive appeal but also because of the ways in which it is able to quietly speak to these wider concerns. However, to say that criminal commodities move is not the same as saying that they constitute flows. The metaphor here obscures some more complex empirical realities that challenge the discursive associations that "flow" evokes. The idea of flow suggests movements that are uniform, continuous, and smooth. One thinks, perhaps, of a mature river channel. The commodity mobilities associated with transnational criminal markets have anything but these qualities. These movements are not flows of a single commodity. They are not uniform. Rather, illicit commodity movements are often highly diverse. They might, for example, include multiple illicit or illegal commodities or mixtures of many illicit, illegal, and licit goods within a single "flow."

> Popular media would have it that drugs travel a "drug route," arms an "arms route" and computers a more cosmopolitan "high-tech corridor." In fact, shipping routes are markets, and thus they are matters of opportunity. . . . Once routes are operating with confidence, all manner of goods can pass along them. A shipping container can "contain" arms, cigarettes, and the latest pirated DVDs, along with a host of other commodities ranging from the seriously illegal to the merely mundane. In fact, such transits work more smoothly than they would if all routes were separate. (Nordstrom, 2007: 8)

Criminal mobilities can be thought of as being multiple and complex in other ways that exceed their rendition through a simplistic metaphor of flow. Kilcullen, for example, has identified the nesting of illicit movements within other, licit mobilities. He cites the investigative journalist Matt Potter (2001) who has discussed weapons smuggling mobilities that operate within humanitarian networks around the Horn of Africa, recognizing the problems that this creates for attempts to shut these illegal smuggling networks down.

> Some (though, of course, by no means all) of the same air charter companies that operate humanitarian assistance flights into drought-striken or conflict-affected areas such as the Horn of Africa also smuggle weapons, drugs and other contraband. Humanitarian aid workers and NGOs are perfectly well aware of this, but neither they nor the governments involved in relief efforts can shut down these trafficking flows, since it would mean an end to the movement of humanitarian assistance cargo. (Kilcullen, 2013: 113)

We have here, then, complex movements within other complex movements. The nature of these mobilities, therefore, is not captured by the notion of flow, which is able to convey little, if anything, of the embedded nature of these mobilities. In not speaking of the contexts within which criminal mobilities are nested or embedded, the idea of flow has the effect of directing attention away from these contexts and focusing it solely on mythically simplistic illicit flows in a way that renders them disembodied, despite considerable empirical evidence to the contrary. Any response that does not acknowledge the other mobile contexts that facilitate these movements and within which they are nested is likely to offer, at best, only partial solutions to the issues they generate.

As noted above, the use of the metaphor of flow to describe the movement of illicit commodities shows distinctive historical trajectories (Hall, 2017). Figure 6.1 is a graph analyzing the use of the term "drug flow" and a host of near variations of this term, in digitized Google books. It shows that the flow metaphor first appeared in the late 19th century but was relatively little seen until the 1970s. Subsequently, there was a steep rise in its use up until the year 2000, followed by a slackening off of its usage. The earliest use of the term "flow" in association with the movement of drugs was largely used to refer, in a medical sense, to the movement of drugs *through* the human body, a use to which the metaphor is empirically appropriate. However, subsequently this usage was extended in its becoming widely deployed in the description of the movements of drugs *outside* the body. The first detected instances of the use of the flow metaphor to refer to the movement of drugs across geographical space was from the 1930s in the proceedings of a host of U.S. congressional and legal hearings, and occasionally academic publications. Thereafter the flow metaphor seems to have been largely restricted to legislative, political, and academic circles until the 1990s and 2000s when it appeared to enter, albeit relatively slowly, more widespread usage. The deployment of the flow metaphor with the annual United Nations Office on Drugs and Crime's *World Drugs Report* (and its antecedent *Global Illicit Drug Trends*) seems to show a similar trajectory, reaching a peak around 2010 from a low initial recording in the late 1990s and early 2000s (see Table 6.1). While this analysis is nothing more than exploratory and indicative, it does point to a, potentially politicized, way of describing the movement of an illicit commodity, emerging across the temporal course of that commodity being constructed as an existential and security threat at the heart of an Americanized worldview of organized crime that has prevailed unchallenged until very recently (Collins, 2014b; Woodiwiss, 2003; Woodiwiss and Hobbs, 2009).

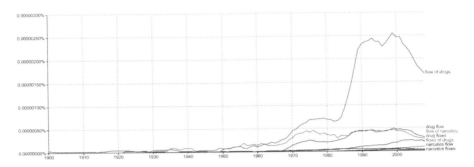

FIGURE 6.1. Analysis of the appearance of the terms "drug flow," "drug flows," "flow of drugs," "flows of drugs," "narcotics flow," "narcotics flows," "flow of narcotics," and "flows of narcotics" in digitized Google books.

The singular readings of illicit and illegal movements that tend to prevail within official, and many popular, accounts are significant, for they render these mobilities subject to myths evoking both existential threat (Woodiwiss and Hobbs, 2009: 124) and the possibilities of eradication. In presenting them as singular uniform, disembodied movements, rather than as the messy, embedded mobilities that, in the majority of cases, they are, they present them as phenomena that are more readily amenable to universal policy responses that isolate them as operating outside of mainstream economic and social processes. Such representations are part of the process in which policy responses to illicit and illegal economic activities are turned outward toward a somewhat mythical terrain from where the illicit is imagined to originate, rather than inward toward the operations of the economic mainstream.

Furthermore, empirically these flows are not continuous or smooth. Rather, ethnographic studies have shown them to be discontinuous, irregular, vulnerable, and often very slow (Bhattacharyya, 2005, 4; Brooks, 2012; Nordstrom, 2007; van Schendel and Abraham, 2005: 48). Looking inside these mobilities, as these various works do, disrupts the assumptions attached to prevailing understandings of illicit commodity movements as "flows." Rather, under these critical lenses, they appear very un-flow-like. The idea of criminal mobilities as flows is also inattentive to the restricted agencies of those involved in moving illicit commodities. As Reuter (2014: 34) reminds us, the agency of these actors is far from perfect, their environmental and market knowledges partial, and their responses

TABLE 6.1. Analysis of the Frequency of Occurrence of the Term "Drug Flow" (and Associated Terms) to Refer to the Movement of Illicit Narcotics across Space, in the United Nations Office on Drugs and Crime Annual Reports *Global Illicit Drug Trends* and *World Drugs Report*

1999[a]	4
2000[a]	4
2000	12
2001[a]	10
2002[a]	8
2003[a]	3
2004	6
2005	9
2006	10
2007	41
2008	11
2009	25
2010	94
2011	55
2012[b]	13
2013[b]	17
2014	19
2015	31

[a]*Global Illicit Drug Trends*.
[b]Shorter version of the report with statistical information presented separately.

to changing market conditions, for example, in terms of shifting trafficking routes following law enforcement successes, frequently slow. If a flow metaphor were to work in capturing more of the empirical realities of the movements it describes, it would be better built around evocations of rills, bifurcation, and ephemerality than that of the powerful and stately progress of a mature river channel.

Moving beyond instances of the deployment of specific metaphors and cartographic motifs, it is possible to discern ways in which official texts speak of the nature of the phenomenon they describe in ways that reinforce perspectives of their institutional authors. The categorization of phenomenon within official reports is not a reflection of any natural, external order, but is rather a discursive strategy that imposes orders across complex, diverse, and discrete empirical terrains. The gathering together, in this way, of discrete phenomena and their representation as singular or interrelated processes or issues has long been recognized as a tactic deployed across swathes of media and political discourse (Edelman, 1988; Hay, 1996). Specifically, the elision of discrete phenomena associated with criminal or illicit markets and, through a range of institutional and discursive strategies, the imagination of coherent narratives of threat under the banners of organized and global crime has been highlighted within a number of critical criminologies (Hobbs, 1998, 2013: 18; Levi, 2009; Woodiwiss and Hobbs, 2009). Such rhetorical elisions, and equally the exclusion of other, empirically similar, phenomena from these discourses of organized and global crime (Hall, 2013: 370; Ruggerio, 2009: 119), are important in the maintenance of particular policy positions and discursive mappings of plural global economic terrains. For example, Woodiwiss and Hobbs (2009) have argued how such narratives have been crucial in mobilizing specifically American hegemonic mappings of organized crime and achieving significant degrees of international consensus through transnational institutions such as the United Nations and the intergovernmental political forum of the world's most highly industrialized democratic nations, the G-7/8, as well as national government departments, criminal justice groups, and investigative agencies (see also Hobbs 2013: 18):

> We do not attempt to deny the existence of drug and people trafficking, and myriad other acts defined as criminal by the legislative process.... However, loading these activities into a single category marked "organized crime" creates a sense of threat that is far more ominous and substantial than is warranted after realistic analysis. (Woodiwiss and Hobbs, 2009: 124)

Reports that emanate from these official sources can be read through this critical lens as, frequently, attempts, which routinely enter media, political, and policy discourse, to achieve this "loading of discrete phenomenon into a single category" in ways that accord with wider institutional, political, and policy worldviews. Documents, often produced on annual or regular cycles, such as the United Nations Office on Drugs and Crime's *World Drug Report* and a host of equivalents such as the *Global Report on Trafficking in Persons* (2014) can be interpreted as attempts to elide a host of discrete phenomena grounded in specific spaces, contexts, and processes and present them as part of singular "global" processes or problems, which consequently point toward international, universal, responses rather than those grounded in or tailored toward the particularities of issues within discrete spaces. The visual representations of these issues and particularly the visualization of transnational connections and movements play significant roles in the imagination of global organized crime problems within these documents.

One of the clearest examples of this process is the construction of a global or world drugs problem through the publication of the United Nations Office on Drugs and Crime's annual *World Drug Report*. As Inkster and Comolli (2012: 16) point out, "The UNODC itself is cautious about the limitations of its approach, pointing out for example that its annual reports on what it calls the 'world drug problem' describe a problem for which there is no internationally agreed definition." *World Drug Reports* are comprehensive assessments of international trends in the production, trafficking, and consumption of illegal drugs. The reports draw together a range of national and international data and include extensive deployment of maps, graphs, and tables along with detailed textual commentary. The *World Drug Report*, then, is both a lens through which international drug trends are viewed and a means of constructing discursive understandings of these trends. The first *World Drug Report* appeared in 1997 and has been produced every year except 1998 since then. In some early years it appeared under the title *Global Illicit Drug Trends* (*www.unodc.org/wdr2015/en/previous-reports.html*). The issues that these reports discuss are presented explicitly as a series of global problems. For example, this global discourse is apparent in the following quotation, which is drawn from the opening of the executive summary of the *World Drug Report 2015* (2015c: ix):

> The World Drug Report presents a comprehensive annual overview of the latest developments in the world's illicit drug markets by focusing on the production of, trafficking in and consumption of the main illicit drug types and their related health consequences. Chapter 1 of the World Drug Report 2015 provides a global overview of the supply of and demand for opiates, cocaine, cannabis, amphetamine-type stimulants (ATS) and new psychoactive substances (NPS), as well as their impact on health, and reviews the scientific evidence on approaches to drug use prevention and addresses general principles for effective responses to treatment for drug use.

Such presentations represent rhetorical elisions in two senses. First, they draw together and present within a singular discursive frame discussions of very different drugs that have, as the report itself demonstrates, very different geographies of production, trafficking, and consumption; very different psychoactive properties and effects; and that circulate within very different regional and market contexts, often peopled by very different actors. Second, the reports elide the drug problems of specific regions, presenting them as constitutive of a singular global drugs problem. However, empirical readings of the drugs economies of specific regions reveal them not to be part of a singular global problem, but rather the result, in part at least, of sets of geographically specific location factors (Brophy, 2008; Castells, 2000; Chouvy, 2009; Ferragut, 2012; Goodhand, 2009; Hastings, 2015; Kenney, 2007; Watt and Zepeda, 2012), notwithstanding the roles of various macroprocesses and their articulation with other spaces through series of networked connections (Hall, 2010a). Thus, it can be plausibly argued that the world does not face a singular drugs problem but rather a series of smaller, specific, regional drugs problems. However, the version of, in this case, the drugs problem that is promulgated through these and other official accounts accords well with American-led policy responses, which have dominated the international responses to the question of drugs, and of organized crime more generally, but which are coming under increasing critical and political pressure (LSE Expert Group on the Economics of Drug Policy, 2014; Woodiwiss and Hobbs, 2009).

The "globalness" of the issues discussed in the *World Drug Reports* is constructed in multiple ways beyond its assertion within the textual discussion. These include the frequent presentation of data at a global level throughout the report in the forms of tables, graphs, and graphics (see, e.g., Figures 2.1 and 6.2).

The issue here is not that any of the information being presented in these visuals is inaccurate, in as much as we can be sure of the data associated with illegal drugs, but rather that its presentation at the global scale prompts a preferred reading of these issues as a singular global phenomenon rather than particular regional ones and ones associated with specific drugs. The uniform visual language of these graphics throughout the reports (e.g., colors, fonts, styles) and their embedding within a textual narrative and document structure emphasizing the global nature of the issue under discussion, further reinforces the impression that they speak of a singular global problem. The role of global maps, and the mobilities that are mapped out across them, is particularly important here also. The choice of maps and the global scale strongly suggests that issues explored exist at this global scale. The "surfeit of arrows" (van Schendel, 2005: 41) that

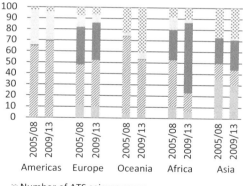

FIGURE 6.2. Distribution of global seizures by drug and region (number of cases) 2005–2008 and 2009–2013. Data from the United Nations Office on Drugs and Crime, response to an annual report questionnaire and other official sources. From United Nations Office on Drugs and Crime (2015c: 37). Reprinted by permission.

represent the movements of illegal drugs across space are typically uniform in color despite their emanating from and moving to, different regions of the world, despite their being moved in very different ways in different regions. They suggest that these flows are more singular and uniform than is the case empirically. The only quality of these movements that is sometimes reflected in these visual representations are the volumes of commodity being moved. Thicker arrows are sometimes used to show bigger movements. The movement of illicit commodities of all kinds, especially across national boundaries, has long been used to generate social and political anxiety (Aas, 2007; Hobbs, 2013; Nordstrom, 2007; van Schendel, 2005; Woodiwiss and Hobbs, 2009). While certainly not intending to deny the realities of illicit transnational commodity movements of all kinds, their presentation in these ways speaks to those anxieties rooted in global crime mythologies by suggesting a far more structurally coherent single global drugs system than appears to actually exist in reality. Returning to Inkster and Comolli's (2012: 16) critical point noted above, while the United Nations Office on Drugs and Crime reports acknowledge the lack of an agreed definition of the "world drug problem," it could be argued that they do a great deal to promote a particularized understanding of it.

Summary

This chapter has explored a number of aspects of illicit and illegal mobilities that suggest further critical analysis of these phenomena at both the empirical and discursive levels and that raise questions of extant and future policy responses. The chapter has argued, for example, that illicit and illegal mobilities are part of complex and multiple networked forms whose spatialities are shaped by a variety of factors; that they are embedded within the spaces through which they move, their societies, infrastructures, and markets; that they "touch down" at all points; that these mobilities are transformative within these spaces and that their effects can be developmentally positive as well as negative, predatory, violent, and destructive; that they are not ontologically distinct from a host of other mobilities characteristic of the contemporary economic and political mainstream, despite prevailing representation to the contrary; and that they are commonly discursively deployed as a way to attach senses of "global-ness" and structural coherence to disparate and discrete empirical phenomena.

These observations, while not necessarily startlingly original insights regarding these mobilities, have tended to be, to date, dispersed across a variety of multidisciplinary critical, methodological, and empirical literatures. This chapter has aimed to draw these insights together somewhat and to advance a view of illicit and illegal mobilities that foregrounds qualities of them and their representation that suggest four further responses. First, these insights suggest a need for additional empirical research. Instances of research where the network connection or specific mobility is the focus of analysis and that explores both the nature of its embedding within the spaces through which it passes and the effects and reciprocities between this mobility and these spaces, are relatively sparse thus far, a limitation that is exacerbated by the diffuse disciplinary locations of this research. GPN and "follow-the-thing" approaches here seem to offer some potential to address this lacuna and have been deployed effectively in some instances. Further research in this and other veins exploring a number of other illicit and illegal mobilities would seem to offer the potential to deepen and extend knowledge of these issues, processes, and their effects.

Second, there appears to be potential to extend and apply insights from work within the critical cartographies tradition to the representation and discussion of criminal mobilities. This chapter has sought to question aspects of the prevailing representations of these mobilities and the wider discursive understandings of illicit and illegal transnational markets into which they are inserted. There would appear to be, then, the potential to explore alternative discursive and cartographic practices to those that have prevailed and gained prominence and traction to date, and that are better able to capture the empirical realities of the phenomena that they represent, specifically the complex, embedded, and diverse nature of specific mobilities and the contingent natures of the criminal markets of which they are part.

This chapter has also pointed to some limitations in extant policy responses to criminal mobilities and organized and transnational criminal markets generally. Thus, third, this chapter has reiterated some concerns regarding the efficacy of universal policy responses. In line with the broader arguments outlined in this book, it has advocated for more geographically specific policy responses to organized and transnational criminal markets and their associated mobilities.

Fourth, this chapter has argued that the embedded nature of illicit and illegal mobilities poses significant challenges to tackling these mobilities and is a quality that is not currently captured within extant policy responses. As Kilcullen argues, "It will be virtually impossible to target a dark network without also harming the community within which it nests" (2013: 112). The chapter has argued that policy responses should not suppose that the illicit and the illegal mobilities discussed here operate beyond the social, economic, and political mainstreams, but rather are deeply embedded within elements of them. Thus, policies should not be oriented outward beyond these realms, toward the somewhat mythical terrains within which organized and global crimes are typically located and seen to originate (Woodiwiss and Hobbs, 2009: 124), but rather inward in attempts to address the multiple ways in which these social, economic, and political mainstreams "domesticate" (van Schendel, 2005: 55) criminal mobilities of various kinds. These issues, among others, are ones that will be considered in greater depth in the following chapter, which seeks to map the application of the insights of this book in thinking critically about how we might effectively respond to the challenges raised by organized crime.

CHAPTER 7

RESPONDING TO ORGANIZED CRIME

INTRODUCTION

There is little doubt that attempts by international organizations, national governments, law enforcement departments, and security agencies to address the problems that organized crime raises cannot be regarded as successful. While there have been many specific anti-organized crime successes, there is no evidence, in a macroeconomic sense, that the economies associated with organized criminal markets have contracted as a result of attempts to tackle them (Mejia and Restrepo, 2014: 28; Reuter, 2014: 40). Indeed, the evidence would seem to point toward the expansion of these criminal markets, and the threats they pose, since the fall of the Berlin Wall in 1989 (Bhattacharyya, 2005; Castells, 2000; Galeotti, 2005a; Glenny, 2008; Madsen, 2009; Nordstrom, 2007). There is a broad consensus across a wide critical literature that current measures to tackle organized crime require significant reform if they are to be more effective (Gill, 2006; Hobbs and Antonopoulos, 2013; Levi, 2014; LSE Expert Group on the Economics of Drug Policy, 2014; Midgley et al., 2014; Varese, 2011: 195–198), but there is only very limited agreement about the nature of this reform (Collins, 2014a; Felbab-Brown, 2014: 41; Midgley et al., 2014) and little direct engagement from the critical academic community around how this might practically be enacted.

We can recognize five fundamental reasons that have underpinned this lack of success. First, as this book has already recognized, anti-organized crime legislation has always created opportunities for criminal entrepreneurs and for expanded criminal markets. Examples of this include the universal prohibition of narcotics, creating markets of extremely high value controlled exclusively by criminal entrepreneurs that have been discussed at many points throughout this book (Caulkins, 2014: 18; Collins, 2014b: 10; Glenny, 2008: 260); the hardening of international borders that have created markets for people traffickers who promise to get people across these international borders illegally; and the banning of the e-waste trade between developed and developing countries, which created an illicit e-waste trade niche that has been exploited with alacrity by criminal networks (Hudson, 2014: 778).

Second, the challenges of tackling organized criminal groups are significantly heightened by their having proven to be both mobile, although far from weightlessly so (Reuter, 2014: 37–39; Varese, 2011), and adaptive (Midgley et al., 2014: 41). Thus, anti-organized crime initiatives, particularly if successful in terms of disrupting the specific criminal economies they are aimed at, are not without consequence, many of which are regularly not foreseen by those devising and enacting policies and operations. Many authors have recognized the adaptive mobility of organized criminal groups and economies with respect to location, victims, or specific criminal activities in response to successful anti-organized crime initiatives or changes in socioeconomic and/or political conditions within their regional or national settings (Moynagh and Worsley, 2008: 179; Varese, 2011: 196; Weinstein, 2008). The unforeseen costs of anti-organized crime policies and initiatives include, for example, the creation of internally displaced populations (Atuesta Becerra, 2014), increases in violence and structural changes within these economies, and the disruption of the developmental or stabilizing effects of organized criminal groups within otherwise unstable regional contexts or those remote from formal economic growth and political governance (Bhattacharyya, 2005: 118; Daudelin, 2010: 15; Felbab-Brown, 2014: 46; Goodhand, 2009; Hall, 2010b: 844; Hudson, 2014: 790; Lee, 2008: 345; Nordstrom, 2007; Varese, 2011: 198; Watt and Zepeda, 2012: 223-224); or the shifting of criminal economies or mobilities to alternative spaces (McCoy, 2004; Moynagh and Worsley, 2008: 181). For example, on the very day that this introduction was completed, a number of British newspapers reported the findings of a U.K. Parliament, House of Lords committee on the unintended consequences of Operation Sophia, a European Union anti-people smuggling operation in the

Mediterranean Sea. The operation was intended to disrupt people smuggling from Libya into Europe by intercepting and subsequently destroying large ships used to smuggle migrants. While the operation made no significant impact on the numbers of people attempting to enter Europe illegally by boat across the Mediterranean Sea—indeed, the trade seems to have increased since the operation began—it also had the unintended consequence of making the illegal migration process more dangerous, resulting in an increase in migrant deaths. Following the interception of large ships as part of Operation Sophia, people smugglers adapted and started using much smaller craft, often inflatable dinghies, which are significantly more vulnerable in rough seas than larger ships (Foster, 2017; Haynes, 2017). While the report of the House of Lords committee, to which these articles refer, offers only an initial assessment of Operation Sophia, if the findings are proven accurate, they will fit into a long tradition of anti-organized crime initiatives that have been undermined by unintended and unforeseen consequences that have stemmed from the adaptability or mobility of organized criminals. Ultimately, it would seem that the question of addressing the problems of organized crime is more one of management than eradication (Collins, 2014a: 8; Gill, 2006).

Third, as was acknowledged in Chapter 3, robust data and evidence, upon which responses to organized crime may be based and their effectiveness assessed, are extremely underdeveloped and many of the assumptions that underpin prevailing responses to organized crime remain largely untested (Holmes, 2016: 42; Levi, 2014: 8; Midgely et al., 2014: 3, 18, 41; Watt and Zepeda, 2012: 4–5). Until the imperative for reliable evidence is addressed, all discussions about the efficacy of anti-organized crime initiatives are likely to remain somewhat specious.

Fourth, tackling organized crime can be attempted at a variety of different scales from the locally specific contexts within which criminal actors and markets operate and through regional, national, and global contexts that may facilitate these economies (Aas, 2007; Felbab-Brown, 2014: 44; Hobbs, 1998; Hobbs and Dunnighan, 1998; Hudson, 2014: 780; Levi, 2014: 12; Midgley et al., 10; Passas, 2001). This raises a series of challenges including at what scale, or scales, to direct anti-organized crime policies and initiatives; the often radically different approaches, and consequently the associated agencies and institutions involved, that these various scales demand; and how to articulate anti-organized crime responses across different spatial scales (see Figure 7.1).

Scale	Element of transnational organized crime system	Theory of change
Local	"The business"	Deterrence
Local/regional	Facilitation networks	Deterrence, severing the links between politics, the state and crime, managed adaptation
Local/regional/national	Support networks	Managed adaptation, cultural change, economic transformation
Regional/national	Social, political, and economic vulnerabilities	Managed adaptation, cultural change, economic transformation
Global	Global political and economic vulnerabilities	Managed adaptation, global regulation

FIGURE 7.1. Theories of change with transnational organized crime system. Based on Midgley et al. (2014: 10).

A number of commentators have noted, for example, that many approaches to tackling organized crime, particularly deterrence approaches based around the prosecution of criminal actors, do nothing to address the conditions that give rise to organized criminal economies (Jesperson, 2014: 151; Varese, 2011: 195; Watt and Zepeda, 2012). This points to the challenge of constructing anti-organized crime programs that address multiple scales, something that has proven difficult to enact in practice. Midgley and colleagues (2014: 10) note, for example, that "these [anti-organized crime] efforts have not been evenly distributed." This has led to a lack of balance with the majority of anti-organized crime policies and initiatives designed to suppress the supply of illicit goods and services, while few serious attempts have been made to address the great public demand for these goods and services that exists, particularly in the global North (Bhattacharyya, 2005: 159, 174; Collins, 2014a: 8–9; Moynagh and Worsley, 2008: 179; Scott-Clark and Levy, 2008). Similarly, it has been recognized that as well as this scalar multiplicity these approaches might span numerous, discrete fields of practice. Glenny (2011: 269), for example, has been critical of the overreliance on technological solutions to cybercrime and has bemoaned the lack of dialogue with and research into the psychological motivations of those active in cybercrime economies.

Finally, throughout the history of anti-organized crime policy, its usefulness as a tool for governments to achieve their desired political aims has trumped the imperatives suggested by evidence-based policy avenues. The politicization of anti-organized crime policy, then, is nothing new and dates back at least to the opium wars in the mid-19th century, when the British government sought to protect the significant revenues they derived for their Indian colony through illegal opium exports to China, a trade that had been going on since 1773 (Knepper, 2009, 2011; McCoy, 2004: 35–39; Woodiwiss and Hobbs, 2009). That evidence-based policy has gained such limited traction within anti-organized crime policy circles should, perhaps, come as little surprise, given the roles that ideology has evidently played in shaping their agendas. The politicization of anti-organized crime policy and initiatives is responsive to both short-term political expediencies (Levi, 2014: 11; Mejia and Restrepo, 2014: 32) and to the projection of more extensive ideological projects (Bhattacharyya, 2005; Gregory, 2011; Watt and Zepeda, 2012; Wilson, 2009).

This chapter will explore in more detail some of the challenges that tackling organized crime poses. It will do this with reference to some examples of extant approaches. It will begin by looking at some of the factors that dominate anti-organized crime initiatives at present, namely, prohibition, supply suppression, altering the regional contexts within which organized criminal economies operate, achieving global regulatory approaches, and arguments that anti-organized crime policy and practices have been deployed in ways that are governmentally repressive. Inevitably this discussion will heavily reference the United States-led "War on Drugs." The chapter will then turn to consider more liberal, permissive alternatives, their contributions and limitations, and the emergence of more plural policy narratives. It will conclude by returning to some insights raised throughout this book and consider how an economic geography perspective, specifically, might contribute to these future policy and practice trajectories.

EXTANT RESPONSES TO ORGANIZED CRIME

The United States-led "War on Drugs," a term that originated in the early 1970s under the presidency of Richard M. Nixon, although the approach it referred to predates this administration, is an approach to addressing organized crime that has attained a hegemonic status internationally in anti-organized crime politics, policy, and practice circles (Aas, 2007: 121; Inkster and Comolli, 2012: 9). The history of this policy, enacted through three international conventions, the Single Convention on Narcotic Drugs (1961), the Convention on Psychotropic Substances (1971), and the United Nations Convention Against Illicit Traffic in Narcotic Drugs and Psychotropic Substances (1988) (Inkster and Comolli, 2012: 149), is readily available from a host of sources and will not be rehearsed here. Suffice it to say that this approach to tackling organized crime is built around the principles of prohibition of the illicit, suppression of supply, militarization, universality, and global governance. These are principles that to varying degrees permeate wider approaches to addressing the problems of organized crime internationally (Jakobi, 2013; Midgley et al., 2014: 35–38; Sharman, 2011: 3; Varese, 2011: 27). The "War on Drugs" has generated an extensive critical literature and calls for major reform from both within and beyond the traditional constituency of critical scholars have been growing since the mid-2000s (Bhattacharyya, 2005; Collins, 2014a: 9, 2014b: 6; Glenny, 2008: 260; Inkster and Comolli, 2012; LSE Expert Group on the Economics of Drug Policy, 2014; Watt and Zepeda, 2011: 227), while some global states have in recent years experimented with limited reclassification, decriminalization, or legalization of some narcotics. These calls for reform, though, are not universal and there has been only a limited shift in the international political drug policy discourse (Felbab-Brown, 2014: 41).

The numerous criticisms that have been leveled at the "War on Drugs" highlight not just the specific flaws of this policy, but more general concerns about the ways in which anti-organized crime policy and initiatives have evolved and been enacted since the late 20th century. These include the flaws of supply-side policies, including the displacement effects of successful interventions; the uneven geographical impacts of universal policies; and the failure to attend to the conditions that gave rise to and sustain criminal economies and governance through crime and the deployment of anti-organized crime policy and practice as the projection of a form of repressive internal governance and external geopolitics.

The overwhelming policy response to economies based on the transnational distribution of illicit commodities has largely been based on the prohibition of their consumption and the suppression of supply through a variety of methods such as military intervention, the destruction of production sites, and the heightening of border security (Collins, 2014a: 10). This response is based on the intuitively appealing assumption that by controlling, or even better eradicating, the supply of these illicit goods, then their consumption will be significantly reduced. This is achieved, it is argued, either through making these goods difficult to obtain or by cutting supply to such an extent that they are only available to potential consumers at prohibitively high prices (Collins, 2014b: 8). The evidence of growth within a number of these markets over the periods of these policy regimes, however, does not support the logic that has underpinned them (Castells, 2000; Felbab-Brown, 2014: 41–42; Galeotti, 2005a: 1; Madsen, 2009; Mejia and Restrepo, 2014: 31; Nordstrom, 2007; 2011; Phillips, 2005; Reuter, 2014: 40; Wilson, 2009).

The reasons for the failure of these approaches are multiple and are likely to be regionally and market-contingent. In some instances, it is clear that a lack of formal territorial governance makes any official international policy stance practically irrelevant. If, for example, a proportion of a territory lies beyond the control of its own national government and international authority, it does not really matter what the official stance toward the illicit commodities produced there is, because it is unlikely to impact upon these spaces, subject, as they are, to alternative governances that are likely characterized by a high presence of criminal and illicit actors (Chiodelli, Hall, and Hudson, 2017). In Colombia a lack of effective state governance coupled with the few economic alternatives available to the population within the licit economy have been cited as underpinning the failure of coca crop eradication efforts (Felbab-Brown, 2014: 43). Thus, regional governance deficits have been noted as one of the factors underpinning the geographies of the production of illicit commodities in a number of different contexts (Abraham and van Schendel, 2005; Aning, 2007; Bagley, 2005; Brophy, 2005; Costa and Schulmeister, 2005; Goodhand, 2009; Hignett, 2005; Kupatadze, 2007). The case of Colombia above reminds us of the point made in Chapter 5, though, that typically these factors are found in combination. This points to the necessity of policy and practice that is alive to this locational multiplicity.

In other cases, the efficacy of these international efforts has been compromised locally by corruption that has effectively shielded specific spaces from these policy-related initiatives. Extensive bribery, for example, has been noted as the cause of the Mexican army's ineffectual involvement in extensive drug crop eradication programs there in the 1940s and 1950s (Watt and Zepeda, 2012: 32–33). Others have cited the ability of criminal economies to be mobile with regard to location as a means of evading interdiction efforts. Successful suppression of illicit production or criminal supply chains in one location, the argument goes, will likely see them subsequently emerge elsewhere beyond the range of these actions, a widely observed phenomenon that has been labeled the "balloon effect" (McCoy, 2004; Wainwright, 2016: 277). Famously, the literature has cited the circular movement of cocaine production and processing across Colombia, Peru, Bolivia, Ecuador, and Venezuela as a result of successful suppression initiatives, and U.S. Navy disruption of trafficking routes through the Caribbean as a stimulant to their westward shift into Mexico and Central America and their eastward shift into West Africa, particularly Guinea-Bissau (Bagley, 2005: 38; BBC World Service, 2007; Mejia and Restrepo, 2014: 29; Watt and Zepeda, 2008: 81). Of the former, McCoy argues:

> During the 1990s, drug eradication in the Andes failed to reduce the region's coca production. Just as U.S. and UN crop eradication programs had once pushed opium cultivation back and forth across the Asian opium zone, so U.S. bilateral programs have simply shifted coca cultivation up and down the Andes. At the start of the decade, the United States launched aggressive coca eradication programs in Bolivia and Peru, source of 90 percent of the Andes coca crop. Ten years and several billion aid dollars later, coca production in Colombia had boomed to compensate for losses elsewhere, and the United States felt forced to launch a massive eradication program in the midst of an intractable civil war. (2004: 85)

While these production and transit mobilities undoubtedly reflect to some extent the geographies of successful interdiction, robustly identifying causality has proven elusive. The assumption of causality within discussions of criminal market displacement are typically made in the absence of any direct empirical evidence such as testimony from displaced traffickers (Reuter, 2014: 37–39). Thus, Reuter argues that "the balloon effect can be seen as a simplifying metaphor; after all interdiction is just one contributing factor to the observed shifts of trafficking" (2014: 39).

More generally, it has been recognized that supply-suppression approaches are limited by an "inherent paradox" (Collins, 2014a: 10) that tends, in the absence of complete eradication of supply, to return illicit markets to conditions of equilibrium. Where demand for illicit products and services remains constant—and this inelasticity of demand is at least the condition that has prevailed in illicit consumption markets in the global North in recent decades—the effects of successful supply suppression initiatives are likely to produce conditions of product scarcity relative to unchanging demand, and, consequently, increases in the prices of illicit commodities (Wainwright, 2016: 272–273). These increases in price, though, are likely to act as stimuli for producers to increase supply. If increases in supply are affected through this mechanism, they are likely to lead to drops in price and a return to something like the previous market equilibrium. This would suggest that the impacts of supply-reduction interventions are likely to be temporary and with little significant long-term effect on the operation of illicit markets. A side product of this tendency toward equilibrium, however, is likely to be an increase in violence within these markets as successful supply suppression is likely to remove the least effective, and potentially less violent actors, from the market, while the more powerful agents are likely to progressively colonize greater market shares (Midgley et al., 2014: 12; Wainwright, 2016: 272–273). This is something that has been observed empirically within the drug transit economies of Mexico, among other examples (Brophy, 2008: 258; Felbab-Brown, 2014: 46; Saviano, 2016; Vulliamy, 2010; Watt and Zepeda, 2011: 48–50).

Policies, of any kind, that are based on global agreements, standards, or accords that apply to all regions will not produce uniform outcomes throughout the terrains across which they are projected. This is true for two reasons. First, universal policies articulate with locally contingent conditions on the ground, thereby producing outcomes that are equally geographically patterned and contingent. This dialectical relationship has been widely recognized within the economic geography literature (Coe et al., 2007; Hudson, 2005; Mackinnon and Cumbers, 2007; Sheppard, 2008), but little reflected upon within the context of discussions of anti-organized crime policy (Hobbs, 2013). For example, the global response to the problem of money laundering has been to require countries to sign up and adhere to the Financial Action Task Force's anti-money laundering recommendations or face the risk of blacklisting (Hampton and Levi, 1999; Levi and Reuter, 2006; Madsen, 2009: 112–113; Sharman, 2011). This policy regime has drawn many criticisms concerning its effectiveness, but what also has been recognized is that the requirement for countries to adhere to this set of universal standards has produced highly uneven impacts across global space. Critics have recognized that they impact most heavily on small developing nations, small firms, and marginalized populations (*The Economist*, 2013b: 7; Sharman, 2011: 39) for whom they are ill-matched and inappropriate. Thus, Sharman (2011: 6) can argue that there exists "a lack of fit between the rich-world context for which the [anti-money laundering] policy was first designed to operate and the local conditions that obtain." Returning to the case of the "War on Drugs," a number of authors have recognized that the costs of this policy regime fall disproportionately on drug producer and transit countries in the global South rather than those countries in the global North who are primarily consumer countries. These costs include repressive constitutional reform, internally displaced populations, worsening public health, losses of sovereignty and state legitimacy, and growing poverty, inequality, violence, and corruption (Atuesta Becerra, 2014; Collins, 2014a; Daudelin, 2010; Madrazo Lajous, 2014; Mejia and Restrepo, 2014; Saviano, 2016; Vulliamy, 2010; Watt and Zepeda, 2012).

> Viewed from the perspective of producer and transit countries, prohibitionist drug policies are a transfer of costs of the drug problem from consumer to producer and transit countries, where the latter are pushed to design and implement supply-reduction policies. . . . The low effectiveness and high costs of these policies have led the region to ask for an urgent and evidence-based debate about alternatives to strict prohibitionist drug policies. (Mejia and Restrepo, 2014: 26)

Caulkins (2014: 25) argues that this uneven and inequitable distribution of costs raises some fundamental questions for the operation of extant international drug policies. Either, he argues, narcotics consumer markets should recognize and subordinate their own interests below those of producer and transit economies, something that seems unlikely; they should compensate these countries for the harms that these policies produce—this has been observed in some cases, for example, through the distribution of aid from the United States to Colombia; or it should be concluded that the present policy regime is unsustainable.

This brings us to the second point regarding the geographically contingent impacts of universal policy regimes, namely, that despite policy being underpinned by a set of universal norms and agreements, the ways in which these policies are enacted on the ground varies enormously between regions. Thus, the specific manifestation of the "War on Drugs," for example, depends very much on where one is located. In the United States, it is primarily a means of promoting the systematic incarceration of economically marginalized, largely black, populations active in criminalized inner-city drug markets, whereas in producer and transit countries such as Afghanistan, Mexico, and Colombia, it is more concerned with a series of militarized interventions (Aas, 2007; Goodhand, 2009; Gregory, 2011: 243; Watt and Zepeda, 2012). For these reasons the outcomes of universal anti-organized crime policies in these different spaces vary profoundly.

Traditionally, and it is still largely the case despite growing calls for more broadly conceived approaches, official responses to organized crime have been constructed primarily around law enforcement (Jesperson, 2014: 151; Midgley et al., 2014: 5). Despite individual successes, there is evidence to suggest that these efforts have little extensive deterrent effect in many contexts on the actors operating within criminal economies (Levi, 2014: 12) and that in some cases may, indeed, lead to increases in violence in illicit markets (Midgley et al., 2014: 12). This appears to be particularly the case in regional contexts such as Latin America and West Africa, where corruption compromises law enforcement efforts or deflects them selectively toward the weakest market actors, where there is insufficient legal and institutional capacity within some states, and where criminal groups are relatively well resourced compared to state actors (Midgley et al., 2014: 13). Relying overly on law enforcement efforts offers a particularly limited response in those contexts, typically located in the global South, where there are few, if any, alternative employment and income opportunities available within the legitimate economy (Hobbs, 2013: 23). Here the deterrence effect of possible prosecution remains very low and, crucially, law enforcement does nothing to recognize and address the underlying conditions that sustain criminal economies (Jesperson, 2014: 151; Midgley et al., 2014: 10). There is a growing consensus that for approaches to tackling organized crime to be effective in such contexts they must address the conditions within the regions where organized crime is deeply embedded and is able to flourish (Pham, 2011: 86–89). However, to date there is little agreement on how this might be achieved or any robustly evaluated policy model upon which to build. Furthermore, as noted in Chapter 5, the factors that sustain criminal economies have been identified as encompassing a very broad range of issues including state weakness, governance deficiencies, weak rule of law, facilitative institutional environments, economic marginalization, strategic locations with regard to transnational criminal mobilities, inhospitable or remote physical geographies, weak border security, normative influence of gangsterism, and access to weaponry, among a host of others (Hall, 2010a: 8). Furthermore, these factors are typically found in combination within specific regions and these combinations appear to be specific to location. Articulating with the local economic, political, social, and physical geographies that sustain criminal economies in different regions represents a series of major challenges for nuanced policy responses that have, as yet, been little admitted, let alone addressed, within the practitioner, policy, and academic literatures of organized crime.

More broadly conceived responses to organized crime that do seek to engage with the regional contexts within which illicit economies are sustained have included programs intended to produce cultural, economic, and/or political transformation within these regions and often to strengthen the bonds between communities and the state (Felbab-Brown, 2014: 48; Kilcullen, 2013: 50–51; Lilyblad, 2014; Midgley et al., 2014: 26–33, 42; Watt and Zepeda, 2012: 230). Chapter 5 highlighted the reality that one of the factors that occurs in regions where organized crime is extensive and deeply embedded is the presence of cultures of the illicit and an associated social ambivalence toward state legitimacy, which in turn acts to socially legitimize illicit and criminal economies. This is an element present in many of the specific factors identified in Table 5.1 that have been recognized as sustaining the regional presence of organized criminal groups and have been noted in places including numerous remote borderlands, Afghanistan, Colombia, Georgia, southern Italy, and Mexico (Abraham and van Schendel, 2005; Brophy, 2008: 257; Castells, 2000; Costa and Schulmeister, 2005; Goodhand, 2009; Kilcullen, 2013; Nordstrom, 2007; Slade, 2007: 178), although it has been estimated that two-thirds of territorial-sovereign states possess areas that can be categorized as territories of limited statehood within their borders (Krasner and Risse, 2014; Lilyblad, 2014: 74; Menkhaus, 2007; Risse, 2011: 2–9).

In some contexts the relationship between the state and criminal authority is highly ambiguous and it is difficult, if not impossible, to draw the line between powerful illicit actors and the state (Brophy, 2008; Costa and Schulmeister, 2007; Cribb, 2009: 8; Goodhand, 2009; Hall, 2013: 372–374; Hastings, 2009; Hesse, 2011; Midgley et al., 2014: 47–48; Tilly, 1985; Wilson, 2009). Increasing transparency in these contexts is challenging in the extreme. In many of these settings questions of political legitimacy are inseparable from those of social and cultural formation there. These is considerable debate about the nature of the social and cultural bonds between communities and illicit actors of various kinds, with some arguing that apparent social bonds and cultures of the illicit are in reality largely the product of coercive relations shaped by potential violent criminal actors (Midgley et al., 2014: 26). However, others have recognized complexity to these relationships, even arguing that local populations are far from passive victims of the predatory power of nonstate actors but may be active themselves in the manipulation of these groups. While the threat and the act of violence is undoubtedly part of these relationships, it would appear that it does not exclusively define them (Kilcullen, 2013: 114, 126).

There are a number of ways in which cultural change within regional contexts such as these have been attempted. These have included policing and institutional reform, media campaigns, community development programs, and civic engagement. The evidence for the success of these measures, though, is somewhat mixed. Undoubtedly, there are examples of isolated successes, but individual projects are rarely part of wider development strategies so mobilizing cultural change has proven problematic. Furthermore, the lack of reliable, nuanced data and agreed methodologies has blighted attempts to build a convincing evidence base upon which to assess programs of cultural change against the normative influence of organized criminal groups (Midgley et al., 2014: 27–29). There seems little doubt, in general terms, that enhancing social capital is a route to challenging the social and cultural legitimacy of organized criminal groups (Midgley et al., 2014: 28). However, again the geographical contingencies of these legitimacies must be acknowledged, as must their articulation with, potentially, a host of other regionally specific factors. Translating the general recognition that the enhancement of social capital is a weapon in the fight against organized crime into widespread programs of change has yet to be realized.

Similarly, the equation of economic growth with the marginalization of illicit economies and their associated actors has proven difficult to realize in actuality. This strategy is based on the observation that a lack of legitimate economic opportunities often appears to be a factor in recruitment to illicit alternatives (Aning, 2007; Bagley, 2005; Castells, 2000; Glenny, 2008; Hignett, 2005; Moynagh and Worsley, 2008; Slade, 2007) and that economic marginalization has been cited as a factor in state institutional weakness and limited regulatory legitimacy. Pragmatic responses to this problem include the promotion of economic development and diversification, infrastructure provision, and alternative livelihood strategies, often involving crop substitution programs as alternatives to the cultivation of illegal drug crops. Midgley and colleagues have argued, however, that the efficacy of the assumptions that underpin these approaches is questionable and causation here would appear to be both less clear and more complex. There is, then, little clear evidence pointing to a direct relationship between levels of development per se and organized crime within specific regions.

> Many of the assumptions and consequent interventions that underpin these strategies are not supported by a clear evidence base. TOC [transnational organized crime] is for example often assumed to flourish in contexts where state capacity is weak or absent, and where institutional capacities to limit and prosecute violence are ineffective. As states become more developed economically and stronger institutionally, it is assumed that their capacity to enforce laws and regulate crime within and across their borders will improve. However the evidence would seem to suggest that the very nature of TOC is integrally related to the economic structure of a country. Therefore as economic development takes hold, so the nature of TOC also changes. In some cases this may result in a reduction in the scale and impact on TOC related violence, in others it may lead to an increase in violence, and in yet others it may lead to a displacement of violence to new and possibly far off locations. (Midgley et al., 2014: 29–30)

Given this, it is not surprising to find that the results here are mixed with some individual successes not translating into convincing answers to the general problem of organized criminal presence (Midgley et al., 2014: 30–32). It has been recognized that the promotion of alternative livelihood strategies has the potential to impact upon the scale of illicit economies, potentially to greater extents than eradication-based approaches. However, to date, these strategies have not been sufficiently funded, managed, or sustained. To be successful, such strategies should recognize and engage with all of the drivers of illicit economic activities, something that has rarely been achieved in practice thus far (Felbab-Brown, 2014: 120–121). Inevitably, approaches to organized crime that seek to change the regional contexts within which illicit economies flourish will be multifaceted and multiagency in approach (Jesperson, 2014: 149–150; Levi, 2014: 12; Midgley et al., 2014: 42).

While intuitively appealing, constructing a sound evidence base and identifying causality in the relationships between both income and employment and levels of criminal economic activity has proven elusive (Midgley et al., 2014: 30–33). The most significant barrier within this relationship would seem to be the adaptability of organized criminal groups. Change within the legitimate economy does not seem to cause these groups to simply melt away as illicit markets cease to be the only game in town. Rather, these groups have been observed adapting to their changing economic circumstances and shifting toward the exploitation of alternative opportunities (Midgley et al., 2014: 30–33; Weinstein, 2008) or developing the capacity for more sophisticated forms of organized crime such as cybercrime (Ebbe, 1999; Glenny, 2008, 2011; Wright, 2006). Hastings, for example, discussing the case of Somali maritime piracy, has argued that economic development in this case is unlikely to cure the problem of piracy. Rather, he argues, it is as likely to produce a more sophisticated manifestation of the problem as a stronger state and economy develop the capacity to absorb extensive illicit markets for the sale of stolen ship's cargo, something that has proven impossible within the fragmented political and weak economic contexts that have prevailed in Somalia during the emergence of its maritime piracy economy (2009: 222). Midgley and colleagues (2014: 31) also caution that some of the countries most plagued by the presence of well-developed, endemic, organized crime groups and economies are middle-income nations such as Mexico, Brazil, and Jamaica. Furthermore, economic development can foreshadow increasing levels of consumption of illicit products or services (Midgley et al., 2014: 33). Finally, there is ample evidence that criminal groups are not simply rational economic entities, but rather they are often also characterized by networks of powerful social relations that bind their members together (Albini, 1971; Ianni and Ianni, 1972; Lombardo, 1997; Midgley et al., 2014: 32; Stephenson, 2015). Economic development and liberalization, then, can bring with them unforeseen and potentially surprising effects that may increase or transform, rather than decrease, the levels of organized crime within regions (Midgley et al., 2014: 30–31; Watt and Zepeda, 2012; Weinstein, 2008). Therefore, rather than promoting the economic development of regions per se as a way of tackling organized crime, it is important to account for the wider political economy of these spaces and wider terrains that they are articulated with in multiple ways (Midgley et al., 2014: 32). The promotion of economic development, then, as *the* route to tackling organized crime, does not necessarily offer the panacea it might first appear to do.

As noted at the head of this chapter, it is possible to attempt to address organized crime by engaging it at a number of different scales. In addition to attempting to affect change within nations and regions, a number of responses have attempted to alter the global context. It is well evidenced within the literature that organized criminal groups, and other illicit and liminal actors, exploit uneven regulatory landscapes and especially those spaces or sectors where regulation is lacking, incomplete, or absent (Hudson, 2014; Midgley et al., 2014: 35; Nordstrom, 2007; Urry, 2013; Zook, 2003). The issue of contingent regulatory landscapes has been addressed through calls for, or attempts at providing, global regulations, conventions, and standards through international cooperation and coordination, often involving global bodies such as the United Nations Office on Drugs and Crime. Examples of attempts to produce global regulations, conventions, and standards include the prohibition of narcotics, the negative effects of which have been addressed above; the Kimberley process for the certification of diamonds, an attempt to address the exchange of conflict diamonds (see Wright, 2004, for an extensive discussion of the evolution and implementation of the Kimberley process); the Convention on International Trade in Endangered Species of Wild Fauna and Flora (CITES); and the recommendations of the Financial Action Task Force designed to counter money laundering.

Midgley and colleagues (2014: 36–38) summarize potential limitations associated with attempts to tackle the power of organized criminal groups through global regulation. These include compliance and enforcement issues such as those related to weak national government enforcement, noted above, and the voluntary nature of the majority of global standards; uneven international uptake, undermining the aim of producing smooth regulatory environments; overly complex standards; the time-intensive nature of negotiations (many years in some cases); global regulation inhibiting individual state action against organized crime; criticisms of the conservative nature of some global bodies such as the United Nations; the problems of national agencies sharing key information; and the difficulty in measuring the impact and progress of global regulation, particularly at the national level and below. While the majority of these global initiatives have been primarily aimed at addressing the supply of illicit commodities, they are also intended to have some impact on social attitudes toward their consumption. Midgley and colleagues (2014: 39) also recognize the demand-reduction potentials of high-profile campaigns and against the consumption of illicit products. An example is the 2016 United Nations–led #Wildforlife campaign to end illegal trade in wildlife backed by famous figures including Brazilian fashion model Gisele Bündchen and soccer player Yaya Touré. While there is evidence that such campaigns impact through raising awareness, this evidence is anecdotal and the impacts on demand apparently limited (see also Duthie, 2014). The Kimberley Process Certification Scheme, though, is broadly regarded as a success, despite it being compromised to some extent by weak national government enforcement (Haufler, 2010). The potentials of the Kimberley Process have been recognized as lying in the unusually concentrated nature of the diamond trade, and reservations have been expressed regarding the transferability of this form of global regulation to other illicit and conflict commodities (Haufler, 2010), although this skepticism is not universal and others see the principles that underpinned the Kimberley process informing initiatives related to other industries such as the European Union's initiative on illegal logging and the United Kingdom Prime Minister's Extractive Industries Transparency Initiative (Wright, 2004: 706).

A number of critical commentators have argued that anti-organized crime policies have been deployed either internally or externally as forms of repressive "shadow" governance reflecting U.S. imperialist geopolitical visions and the promotion of associated neoliberal agendas (Aas, 2007; Gregory, 2011; Hobbs and Antonopoulos, 2013; Madrazo Lajous, 2014; Paley, 2015; Taylor et al., 2013; Watt and Zepeda, 2012; Woodiwiss and Hobbs, 2009). Again the United States–led "War on Drugs" is located at the center of such critiques. It has been widely noted, for example, that drug economies in various developing world regions, and drug problems more generally, have been constructed as national security threats to the West, which has been used to justify repressive militarized interventions into the spaces of progressive opposition and social protest in places such as Mexico and Colombia (Gregory, 2011; Taylor, Jasparro, and Mattson, 2013; Watt and Zepeda, 2012). Taylor and colleagues (2013: 424) are able to argue that the "criminalization of drugs and drug certification processes provided a rationale for the U.S. to insert itself militarily into other countries." To this we might add that these have also been used as vehicles for the United States to insert itself politically within the governance landscapes of a number of other countries that have been associated with extensive drug-producing economies. In addition to Watt and Zepeda's (2012) critique and as noted in Chapter 2, there exists a robust literature that has identified U.S.

intelligence agencies as actors who have been active in the facilitation of illicit and illegal trades to these geopolitical ends through their alliances with local criminal entrepreneurs (McCoy, 2004: 31–32, 43–44).

Further to this, Watt and Zepeda (2012: 20, 134–138) discuss instances of the Mexican government using antidrugs initiatives and resources to justify extensive military deployment throughout the country and the suppression of left-wing agrarian and peasant social movements and guerrilla insurgency in regions such as the Chiapas in southern Mexico. Such initiatives firmly position questions of drugs within a military–security nexus rather than, for example, crime or public health contexts, with uneven and repressive effects (Gregory, 2011: 243, 245; Watt and Zepeda, 2012: 204–206; Woodiwiss and Hobbs, 2009: 123). In summary, "US/Mexican policy under the rubric of the Mérida Initiative has a two-pronged approach: to arm and secure unpopular neoliberal policies, investor rights and US geopolitical interests while quashing and punishing dissent and popular protest" (Watt and Zepeda, 2012: 180).

CHANGING THE POLICY ENVIRONMENT

There have been an increasing number of calls for significant rethinks of existing punitive, prohibitionist stances toward the consumption of some illicit commodities and services, most notably drugs and prostitution, in recent years. One of the most prominent strands of this reformist discourse concerns questions of decriminalization and legalization that have become more prominent as the prohibitionist stance toward illicit consumption that has prevailed internationally came under increasing strain from the mid-2000s (Levi, 2014; LSE Expert Group on the Economics of Drug Policy, 2014; Midgley et al., 2014). These arguments typically acknowledge that the eradication of illicit consumption is an unobtainable goal and that degrees of decriminalization would ensure that consumption takes place in more regulated environments rather than in the underground contexts within which it currently occurs. The goal of these alternative policies is more the management of rather than the eradication of illicit markets and the mitigation of their most damaging impacts (Collins, 2014a: 14; Gill, 2006). Such a shift, it is further argued, would challenge the economic hegemony of criminal actors who control these markets and open up the possibility of revenue generation through taxation, which could fund educational, social welfare, and public health initiatives to help those harmed by their involvement in these markets, either as providers or consumers of illicit goods and services, and by indirectly increasing productivity within economies through reductions in prison populations (Levi, 2014: 9).

It goes without saying that such a realignment of political attitudes toward illicit markets is highly controversial (Midgley et al., 2014: 21–23). However, aspects of this discourse have begun to gain political traction with experiments in the decriminalization of the personal consumption of drugs in a number of territories including Portugal in 2001, for example, and legalizing the selling of sex in Sweden in 1999 (Midgley et al., 2014: 39). Increasing international attention has turned to Uruguay, in the context of experimentation in decriminalization of drugs, who decriminalized possession of all drugs in 1974 (Hetzer and Walsh, 2014). While the 21st century has witnessed the emergence of a degree of policy plurality and growing political willingness to experiment with regard to decriminalization and legalization, on a global level the policy environment toward the illicit has changed relatively little in a fundamental sense as yet.

The effects of the decriminalization of illicit consumption that have occurred are yet to provide enough robust evidence to sustain the wider roll-out of more permissive policy attitudes toward the illicit. The majority of evidence that has emerged is derived from a small number of relatively wealthy countries such as the United States, the Netherlands, New Zealand, and Australia, as well as Portugal and Sweden mentioned above. There is little, for example, from less-secure or less-affluent regional contexts that speaks of the transferability of such policy models (Midgley et al., 2014: 23). Some authors have expressed concern that whatever the health or social benefits of decriminalization and legalization, such policies do little if anything to address the security concerns and levels of conflict and violence associated with illicit economies in fragile producer and transit countries (Atuesta Becerra, 2014: 54). On the latter point, universal permissive legislation would open up the possibility of regulated production in global South locations distant to market in much the same ways that the economic geographies of legal economies are characterized by this distinctive international division of labor.

The impacts of legalization on levels of illicit consumption are difficult to predict (Midgley et al., 2014: 24). Research into narcotics consumption, for example, has found that enormous price increases down the value chain are attributable almost entirely to the effects of prohibition. Given this research, it has been argued that it is impossible to rule out the potentials of very large increases in consumption if legalization reduces this price escalation (Caulkins, 2014: 22). This author argues that looking arithmetically at levels of drug dependency that have been prevented by prohibition challenges the view that prohibition is a failure to serve the interests of the United States and other final-market countries.

Furthermore, there are many examples of active, lucrative gray markets in entirely licit commodities such as the smuggling of cigarettes within Europe (Hornsby and Hobbs, 2007). It is likely, therefore, that gray markets in formerly illicit commodities would emerge following any legalization as producers and consumers sought to circumvent taxation regimes (Felbab-Brown, 2014: 48; Wainwright, 2016: 254). Thus, it is unlikely that the "legal status of highly sought after products is unlikely to prevent TOC [transnational organized crime] groups from being able to gain control over these industries" (Midgley et al., 2014: 25; see also Felbab-Brown, 2014: 48). Undoubtedly, legalization would have to be managed carefully to preclude the emergence of gray markets in legalized, taxed commodities.

It has been recognized, however, that legal providers of drugs in contexts where their production and consumption has been legalized have a number of potential advantages over criminal suppliers. These include the ability to benefit from economies of scale that are not available to criminal suppliers, who are required to keep their operations clandestine; reassurances as to the strength and specific ingredients of drug products and their effects, which are likely to be more robustly tested through regulated supply routes; and the ability to innovate ranges of drug products aimed at different markets such as first-time users or those lacking confidence (Wainwright, 2016: 244–250). On the later point, Wainwright describes a tour of the factory of Dixie Elixirs, a large Colorado-based company supplying a range of cannabis products:

> The inside of the Dixie factory looks rather like the crystal-meth laboratory in *Breaking Bad*. But it owes as much to Willy Wonka as Walter White. While some of the white-coated technicians are working with concentrated narcotics, others are stirring vats of molten chocolate or plucking aluminium bottles off an assembly line. . . . Different parts of the factory put together cannabis-infused drinks, in flavours such as Watermelon Cream and Sparkling Pomegranate, and even products such as massage oils. (2016: 250)

Legalization of currently illegal products, then, would represent a fundamentally disruptive change to the legislative context within which criminal actors operate and would undoubtedly pose serious challenges to the sustainability of their operations. There is evidence, cited by Wainwright (2016: 253–260), for example, of the negative impacts of legalization of cannabis in some American states on the revenues of Mexican cartels. As noted earlier, though, organized criminal groups are highly adaptive to changing economic and political circumstances. The movement of organized criminal actors into other, potentially more predatory, economies following legalization, as has been observed in Mexico where cartels have started to traffic people as well as drugs across the Mexico–United States border (Wainwright, 2016: 253–260), cannot be categorically ruled out. Legalization and decriminalization, then, despite these impacts, do not necessarily cut off the income streams of actors previously active in illicit markets and may, in some contexts at least, see increases in levels of violence, harm, and criminal predation as they react by seeking new illegal markets to populate (Felbab-Brown, 2014: 48; Watt and Zepeda, 2012: 231–232). They can, however, offer the potential to disrupt the operations and income streams of criminal actors and are likely to become more central to anti-organized crime initiatives in coming years. Legalization and decriminalization do not offer a complete panacea in themselves, though.

The decriminalization or legalization of illicit economies requires a great deal of regulatory, monitoring, and enforcement effort by the state (Felbab-Brown, 2014: 48). A strong and comprehensive state presence would be a prerequisite if any move toward decriminalizing or legalizing illegal economies were to be effective at all, let alone if it were to prevent the reproduction of extant inequalities associated with the presence of these economies in fragile state contexts. This suggests the importance and challenge of embedding such policies within wider state-building and development efforts (Felbab-Brown, 2014: 47–48). Most obviously, it is unlikely that criminal actors will simply disappear following legalization or seamlessly enter sectors located within the legal economic realm. Programs of legalization, then, might profitably explore how they address the postlegalization mobilities of criminal actors either by anticipating their diversification into other criminal markets, particularly if these are associated with greater violence and predation, or by the development of programs that facilitate their migration into legal economic sectors.

An alternative strand of thinking that has emerged in response to some of the failures and limitations of attempts to forge global policy and regulatory regimes has been growing calls for policy pluralism (Caulkins, 2014: 17; Collins, 2014a: 13–14; Midgley et al., 2014: 42). This shift in thinking recognizes, implicitly at least, that the geographies underpinning the development of illicit economies within specific regions are not universal (Hall, 2010a). Different regions, characterized by different illicit economic geographies, underpinned by ranges of different location factors, might, it is argued, require different approaches and interventions to address their particular problems rather than more singular, universal models. This is a more humble (Collins, 2014a: 14) policy outlook than the strict, universal, prohibitionist stances that have dominated to date, and it has been accompanied by calls for policy experimentation and monitoring, the strengthening of dialogues between policy and research constituencies, and the advocacy of policy on the basis of evidence rather than ideology (Collins, 2014a: 14; Jesperson, 2014: 151; Mejia and Restrepo, 2014: 32). There has also been a recognition that, under more plural policy regimes, the mobility of organized criminality necessitates international policy coordination to avoid interventions generating cross-border externalities as illicit economies become displaced into regions unaffected by such interventions (Collins, 2014a: 14–15; Midgley et al., 2014: 42–43). Speaking of the flaws of singular policy approaches to drug economies, Collins (2014a: 14) outlines principles that might apply more broadly to policies designed to counter illicit economies:

> The failures of the war on drugs and the "drug-free world" strategy shine a light on a more fortuitous response to the question of drugs and drug policy. An effective and rational drug policy should aim to manage drug harms via a multifaceted and evidence-based approach, not a one-size-fits-all, one-dimensional war strategy based on impossible targets. Managing this problem involves incorporating a broad spectrum of policies and indicators and making them work in tandem rather than in opposition to one another.

Perhaps, then, we are witnessing the first signs of the emergence of policy narratives that acknowledge contingency, multiplicity, and the networking of responses to organized crime.

The reviews of extant responses to organized crime outlined above suggest a number of lessons that might profitably inform ongoing debates about anti-organized crime policy and practice. Notwithstanding that there are many different routes to achieving the aims below, some inherent contradictions, and that the operationalization of these principles is extremely challenging, they might be summarized as:

- Develop a more robust evidence base against which anti-organized crime responses can be evaluated and advocate policy on the basis of experimentation, evaluation, and evidence rather than politics or ideology.
- Accept that policy pluralism is a more effective approach to the plural problems associated with illicit economies than a global or "one-size-fits-all" approach.
- Rather than seeking to eradicate illicit and criminal economies, seek to manage them to reduce the harms they generate.
- Address the multiple contexts that sustain organized crime while recognizing both that these contexts will vary by region and that this is likely to change the nature of organized criminal economies rather than eradicate them.
- Recognize that the best way to tackle extensive criminal economies is to seek ways to reduce the demand for the goods and services they deliver.
- Recognize that organized criminal economies can deliver developmentally positive outcomes in the short term, but that this might lead to more destructive outcomes in the longer term.
- Recognize that anti-organized crime interventions can be made at, indeed are likely to require actions at, a number of scales, across a variety of fields of practice, and involve many agencies.
- Recognize that neither full prohibition nor full legalization of the consumption of illicit goods and services offer a panacea, but that there are merits in elements of each of these stances.
- Recognize that anti-organized policies potentially have uneven effects and unintended externalities that impact both across space and different social groups, and that to militate against harms associated with these is likely to require international coordination and the coordination of anti-organized crime measures and wider development goals.
- Consider selective and strategic interventions while seeking to avoid implicit tolerance of those organized criminal groups or economies not targeted.

- Explore policy models that are less bound to, and constrained by, specific territorial units than is currently the case.
- Harmonize policy in ways that ensure different policies work together rather than against each other, as in the criminalization of drug consumption undermining access to health, justice, and welfare services (Collins, 2014a: 14).
- Do not assume that policy models are transferable across radically different regional contexts.
- Give primacy to specific interventions and policies that do not impact negatively upon the wider population (Felbab-Brown, 2014: 120).

Many of the suggestions above have not been heard within debates about responding to organized crime because they lie outside of legislative and policy models that have prevailed to date. These models can be summarized as having been universal rather than plural; seeking the reduction of the supply of, rather than the demand for, illicit commodities and services; based too heavily on ideology rather than evidence; and seeking the unrealistic goal of eradication, rather than the management, of criminal markets. Collectively, the suggestions above offer significant challenges to these traditional models of response to the problems of organized crime. Ultimately, shifting the nature of the responses to organized crime lies with national governments and as much, if not more so, with major international institutions given the transnational nature of contemporary criminal economies. However, given that the central message of the suggestions above is greater plurality, then, a wider range of voices should seek to and be invited to engage with the policy and legislative processes, such as those of civil society, practitioners, and multidisciplinary research communities. It would be naïve, though, to underestimate the challenges of achieving even some of these goals.

GEOGRAPHICAL ISSUES AND POLICY RESPONSES

The chapter now considers what might be added to the lessons sketched out above, based on the insights of the economic geography perspective outlined throughout this book. That geographical perspectives on organized crime and the groups associated with these markets are in their infancy has been obvious and reiterated at a number of points throughout this book. It will be no surprise, therefore, that there are no substantive extant discussions of what a geographically informed and sensitive anti-organized crime policy might look like. This section begins to sketch out some dimensions along which this perspective might be developed. It takes as its point of departure those calls cited above for pluralism and experimentation within the anti-organized crime policy realm (Caulkins, 2014: 17; Collins, 2014a: 13–14; Midgley et al., 2014: 42).

At its most basic, the geographical perspective explored within this book has demonstrated not only that organized crime is distributed unevenly across global space, but that the conditions that sustain it, as well as its impacts, are similarly unevenly distributed. It has been suggested above, given this reality, that universal policy responses to this geographically differentiated criminal landscape are inherently blunt instruments that are likely to produce differential outcomes across space and are unlikely to successfully articulate with the more microgeographies of organized crime. The notion, then, of plural policy responses sits well with the geographical perspective outlined here. Different conditions in place, this book argues, demand different policy responses.

There might profitably be three dimensions along which policy responses to organized crime can be differentiated. Chapter 5 included a discussion, based on analysis of multiple accounts of regions within which organized crime has become extensive and deeply engrained, of the location factors that potentially account for the presence of organized crime within these regions. This discussion drew on location factors approaches that were developed primarily within economic geography. While it is acknowledged that there have been criticisms that have outlined the limitations of location factors approaches, which include the difficulties of convincingly attributing causality to factors identified and accommodating factors from beyond the region (Healey and Ilbery, 1990: 110), this discussion did recognize a number of factors that tend to be present in different combinations within these "mobbed-up" regions (Glenny, 2008). First, then, anti-organized crime policy might seek to articulate with the specific sets of factors that appear to underpin the development of extensive organized criminal economies in different regions. A first step in this endeavor should be a much more comprehensive and systematic analysis of regional accounts of organized criminal markets than has been possible to date (Hall, 2010a). While state weakness emerged as the most commonly cited individual factor within the accounts discussed in Chapter 5 (see Table 5.1), it was far from the only factor within individual case studies and the presence of organized criminal markets was certainly not reducible to this single factor.

The significantly different combinations of factors that appear to be active in helping to develop and sustain organized crime in different regions are illustrated with a couple of examples drawn from this discussion. Aning's (2007) account of criminal networks in Ghana, West Africa, for example, attributes their continued presence to a host of factors that paint Ghana as a weak state with a diminished licit economy and poorly protected borders, within which gangsterism has garnered significant cultural capital, and continues to do so. In contrast, accounts of the cocaine trafficking cartel economy of Mexico (Brophy, 2008; Watt and Zepeda, 2012) also cite state weakness as a contributory factor, but recognize many other factors that, according to Aning's account, are not present, or at least are less significant, in the case of Ghana. These include close links between state government and organized crime, a location that has emerged as strategic in recent decades, regional poverty and inequality, inhospitable and remote environmental terrain, access to large numbers of weapons, technological advances, and the unforeseen effects of prohibition policies. While this suggests that policy perspectives on these two spaces need not necessarily be entirely discrete, they would necessarily demand a degree of tailoring to their very different circumstances.

Second, as this book has stated at a number of points, the factors that sustain organized criminal economies, and illicit economies generally, operate across multiple scales (Hudson, 2014: 780). It would seem sensible, therefore, for policy similarly to seek to articulate with factors operating at multiple scales. It may be, for example, that within a more plural policy regime that discrete policies are directed toward addressing processes and conditions operating at different scales. This book has argued that the contemporary global economy, as it has developed along a primarily neoliberal trajectory, is facilitative of illicit economic actives of multiple kinds in many different ways that are linked largely to its extensive opacities (see Chapter 5). Clearly, following this argument, there emerges a strong case for fundamental reform of this global economic terrain, despite an apparent lack of political will for this effort and a surfeit of corporate hostility. However, while this would be a welcome if, depressingly, somewhat unlikely development, addressing the microspatialities that have been identified as sustaining criminal markets would require policies that go well beyond those directed at the reform of the global economy. Both formulating such multiscalar policies and ensuring they articulate together in complementary ways are significant challenges that the policy and academic communities might contribute to in the future.

Finally, this book has recognized organized crime as a phenomenon that is networked to a significant degree, both in terms of its organizational forms and in terms of the ways in which criminal markets connect transnational spaces (Saviano, 2008). The discussion in Chapter 5 suggested that network ontologies potentially offer innovative ways of exploring the geographies of a wide range of economic phenomena in ways that exceed some of the limitations of more traditional territorial approaches. While network ontologies have generated a wealth of debate and discussion across the social sciences, it should be emphasized that there remains a lack of consensus concerning the application of these approaches, even among some of their keenest advocates (Bosco, 2006; Dicken et al., 2001; Murdoch, 1997, 1998; Sheppard, 2002). However, a number of points emerge from these literatures that might frame the application of network ontologies to the case of anti-organized crime policy. Most fundamental is the position that regions should be considered primarily as networked units. This position necessitates that regions are not viewed solely as bounded, territorial entities but equally as bundles of connections to other, in some cases spatially distant, places. In analytical terms this grants primacy not to the regional territorial unit but, at least equally, to the network. Network ontologies, then, ask us to consider both regions' connections to other places and the conditions that prevail within these distant networked spaces. Here, it is argued, attention should be paid to the connections that bind places together within networks of regions, the specific connections that flow into and out from regions, the nature of these connections, how they emerged, the ways in which they are sustained, and crucially the ways in which they shape the developmental trajectories of regions.

This book has demonstrated that criminal markets are typically both rooted in the conditions within regions and also, often, in the illicit movements passing through them. However, prevailing anti-organized crime policies reflect not this network perspective but, rather, the territorial thinking that underpins contemporary national and international justice systems. Namely, they seek to tackle organized crime by promoting universal responses or policies and initiatives that target specific regions or nations. However, while such policies may be designed with the intention of addressing the conditions that give rise to specific regional criminal markets, they must inherently fail to address the conditions that prevail in other, spatially distant yet closely networked, spaces. Here such policies appear to replicate flaws within primarily militaristic responses to international terrorism (Ettlinger and Bosco, 2004).

In contrast, network thinking calls for policies to be articulated across the various transnational spaces within illicit networks, to address the conditions within these diverse spaces and the connections that bind them together. At the moment the international coordination of policy is something that appears relatively little practiced but that has been advocated elsewhere in the context of managing the international displacement of organized criminal activities after successful regionally focused deterrence responses (Reuter, 2014: 40). The failure, to date, to enact networked policy approaches stems from both the limited impact that network ontologies have had beyond academic social science and the ongoing hegemony of territorial policy models. If we are to develop policies designed to effectively tackle organized crime, it is important that policy frameworks are rethought. This is only likely to occur when network scholars have sought to engage in sustained dialogues with policy agencies beyond the academy and more international policy communities and forums emerge.

Summary

The various responses to organized crime outlined within this discussion have, largely, failed to curb its growth since the collapse of the Soviet Union in the late 1980s. Most high profile among these responses and now the most widely criticized has been the "War on Drugs" (LSE Expert Group on the Economics of Drug Policy, 2014; Watt and Zepeda, 2012). The critical discourses around the extant responses to organized crime and potential alternatives have grown within recent years. There are a wealth of emergent knowledges that outline potential alternatives to the universal, prohibitionist norm that has underpinned anti-organized crime thinking since the 20th century. Central to these are calls for policy pluralism and experimentation (Caulkins, 2014: 17; Collins, 2014a: 13–14; Midgley et al., 2014: 42), to which, this chapter has argued, a geographical perspective might profitably contribute. Geographical articulations of anti-organized crime thinking are very much in their infancy. While this chapter has sketched out some broad potential directions of travel, there is clearly much work to be done on their detailed development, operationalization, and evaluation. However, policy discourses, in some circles at least, seem more open to these interventions now than they have been at many times in the past.

CHAPTER 8

TOWARD ECONOMIC GEOGRAPHIES OF ORGANIZED CRIME

INTRODUCTION

At the end of Chapter 1 this book posed five questions that mapped out the empirical and conceptual terrains that subsequent discussions aimed to traverse. While it did not seek to necessarily answer these questions fully, the book sought, in each case, to summarize relevant scholarship from across a range of literatures, in places drawing this scholarship together for the first time, to push for conceptual advances and to map out agendas for future research. At the end of this book it is hoped that we are a little closer to understanding what the economic geographies of organized crime might look like. Of course, advancing economic geographies of organized crime is far from straightforward as there has rarely been, within the history of this subdiscipline, consensus about what an economic geography of anything might look like. Recently, for example, there have been calls for more ethnographically founded economic geographies, especially of those contexts where extensive informal and illicit economic markets and networks constitute daily realities that are little captured in official statistics (Bhattacharyya, 2005; Goodhand, 2009; Phillips, 2011; Stephenson, 2015; van Schendel and Abraham, 2005). Certainly, these calls for a more ethnographic economic geography provide the potential for common ground across which dialogues might emerge with those critical scholars of, for example, trafficking

(Nordstrom, 2007), local urban illicit markets (Hobbs, 2013), and money laundering (Sharman, 2011) who have long used ethnographic methods in their work. There is scope, then, should the challenge that this book poses to economic geographers be embraced, for multiple economic geographies of organized crime and the illicit to emerge. This eclecticism and the debates it would be likely to generate are to be welcomed. The questions outlined at the end of Chapter 1 will not be returned to and reviewed now. There is much material throughout this book that reflects, explicitly and implicitly, on the issues those questions raised.

RETROSPECT

This book opened with an acknowledgment that geography's failure to engage to any substantive degree with the global illicit economy generally and the markets of organized criminality specifically are significant lacuna in the discipline's literatures, particularly in those of economic and political geography. The discussion in Chapter 1 included a critical reflection on the concept of organized crime and its situation within wider discursive terrains. It was argued that organized crime can be thought of as a series of industries, global in their extent and importance, which, like all economic activities, display distinctive economic geographies, but which have been almost entirely overlooked by geographers thus far, certainly in any systematic way and for which there has been only limited conceptual and empirical insight produced from geographical perspectives. It was also noted that a wealth of highly politicized renditions of organized crime and its actors have become, and remain, central to contemporary understandings of the global illicit, even within some official accounts, but that these must be treated critically and with care. The book then provided a global overview of the illicit economy, noting its extent, broad spatial contours, and key activities including trafficking, counterfeiting, money laundering, and cybercrime, as well as other more mundane illicit trades. In doing so it highlighted a series of issues that are of relevance to economic and political geography including the roles of a range of illicit actors in geographical processes such as regional economic development, governance, and state-making.

The book also discussed the multidisciplinary literatures of the global economies of organized crime, noting their patchy and somewhat partial and underdeveloped engagements with their spatialities.

It was argued that an extended engagement with the economies of organized crime through the lenses offered by economy geography is timely and of benefit to both the literatures of economic geography and to those concerned directly with organized crime. The discussion drew on literature from anthropology, criminology, policing, sociology, and political science, among other disciplines, and provided a conceptual map of the economies of organized crime, ending with a call for a more spatialized and spatially sensitive reading of these illicit economies, something that this volume attempted to at least point toward if not to fully map out.

The methodological difficulties of measuring the economies of organized crime were fully acknowledged and a critical reading of those data that do exist, which are collected and employed by transnational institutions, law enforcement agencies, and government departments, were outlined. These discussions went on to explore the political ends to which measures of the illicit global economy have been put by these organizations. This discussion focused particularly on high-profile, widely circulated reports such as the United Nations Office on Drugs and Crime's annual *World Drugs Report*. These texts were later also read through the lens of critical geopolitics that examined the tropes and forms of representation deployed in these reports. It linked to recent critical work within criminology that has unpicked the role of official narratives in the construction of organized crime and illicit economic activities as external threats to the nation state, a reading that obfuscates their complex realities, often reproducing easily discredited accounts, primarily for political ends.

This discussion also considered the significant difficulties facing researchers attempting to conduct research on the criminal markets associated with organized crime. It noted the threats to researchers from the subjects of their research and the reluctance of official agencies to release data or to cooperate with academic researchers. It noted also the reluctance of researchers to articulate these difficulties within their published research, creating a silence at the heart of this work that is somewhat defensive and unhelpful to those scholars seeking to follow their direction. The discussion talked about the ways in which the realities of research shape the knowledges of organized crime that have been produced. It explored the possibilities available to researchers and the potential for future research practice and, later, the coproduction of research with actors from beyond the academy.

This book has also provided a discussion of the changing nature of illicit enterprises under the conditions of contemporary globalization and the complex interplay between the potentially homogenizing forces of globalization and the enduring significance of place in the organization of organized criminal groups. It argued, drawing on mainstream economic geography theory, that the forms of illicit enterprises, like many of those in the licit economy, reflect the nature of post-Fordist globalization. However, it also argued that despite this reality, and despite the interconnections between the licit and the illicit economic realms noted throughout the book, organized criminal groups, on the whole, do not mirror in form the corporate structures characteristic of the licit economy. Rather, the book argued that there are some significant structural differences between the organization of the licit and the illicit economies. This suggests that the organizational models of the global economy with which we are familiar through the economic geography literature are not universal, and that there are distinctions in the ways in which these economies are organized. The discussion noted that these are contingencies that have encroached little into the literatures of economic geography to date.

Throughout this book, most specifically in Chapters 2, 3, and 5, there has been a concern with the challenges of mapping and understanding the spatialities of the global illicit economy, both empirically and conceptually. This book explored broad mappings of the distribution of organized criminal markets globally, recognizing the limitations of the available data upon which these are based. It focused particularly on a series of regions with particularly high concentrations of organized crime through a discussion that synthesized a series of contrasting case studies. The discussion then moved on to consider theoretical explanations for these spatialities. It provided a critical reading of a series of theoretical approaches to understanding the spatialities of the global illicit economy including location factors accounts, structural accounts, and multiscalar accounts, finally considering the potentials of network ontologies These discussions noted the intuitive appeal of both location factors and structural accounts, while arguing that both provide only partial explanations. It moved toward arguments for the development of networked, multiscalar understandings of organized crime.

This book moved on to consider the networked structure of much of the transnational economies associated with organized crime and the movement, in this case of illicit goods and services, as transformative within the spaces through which they circulate. In doing so, it has taken issue with conventional theoretical accounts such as global supply chain analysis that tend to focus only on those spaces in which supply chains appear to "touch down," creating the misleading impression that these flows are disembodied entities that pass above, rather than actively through, and are grounded in, many spaces. In doing so, it explored accounts and examples of illicit movements that are grounded in the spaces through which they circulate, both being shaped by these contexts and, in turn, shaping them. The discussion was particularly critical of the prevailing cartographic and textual representations of these movements, arguing that they capture little of their empirical realities and perpetuate politicized renditions of these movements.

Finally, the book moved to consider various official responses to the economies of organized crime. It picked up criticisms within the criminological literature that these responses are shaped by a highly politicized representation of the illicit economy as primarily an external threat to nation states. The failures of extant, primarily prohibitionist, responses were noted and a range of potential routes to enhancing these responses were outlined. The discussion argued for more objective conceptualisations of organized crime by those tasked with managing its excesses and for responses that speak to multiple scales, addressing and tying together conditions in diverse spaces that facilitate the development of illicit and illegal economies and the international mobilities of goods and services associated with these economies.

PROSPECT

Writing this book was at times a frustrating process. There is much more to be said about all of the issues discussed in this book, but, as yet, the empirical foundations upon which to extend these discussions have, in many cases, yet to be adequately laid down. At points the arguments in this book butted up against the limits of what was possible to say without the discussion becoming entirely speculative and either conceptually and/or empirically ungrounded. Thus, the book should be seen as an invitation to engage in the making of future economic geographies of organized crime.

Empirically, this book has highlighted much that remains underpursued and several areas where additional research would provide more adequate bases upon which to develop geographical perspectives on organized crime. Analysis in Chapter 5, for example, was based in part on a synthesis of case-study accounts of organized crime from many different regions. This type of analysis has been and remains a staple of analysis of organized criminal economies. Much of these valuable literatures have been produced by criminologists. Undoubtedly, these should continue to be central to criminological understandings of organized criminal markets. However, future accounts might pay more explicit attention to the kinds of location factors identified in Chapter 5. At times these are too implicit to these accounts and could be more profitably foregrounded within them. Furthermore, comparative regional accounts of organized criminal economies, despite the difficulties of constructing them (Hobbs and Antonopoulos, 2014: 2) and their limited tradition within the criminological literature, would speak directly to many of the geographical concerns outlined in this book. These accounts might seek more multiscalar perspectives than have been the case in the past. Certainly, they should seek to transcend the limitations of statist perspectives that are now widely identified and broadly accepted (Aas 2007; Law and Urry, 2007). Furthermore, empirical research that more explicitly examines the articulation of potentially facilitative macroeconomic contexts with the conditions sustaining illicit markets within specific regions would begin to address the recognition of the need for more multiscalar analysis outlined within this book and following earlier calls from economic geographers (Coe et al., 2007: 20). Notwithstanding again the difficulties of undertaking such research, further research, of the kind pioneered by Hobbs (1998, 2004, 2013; Hobbs and Dunnighan, 1998) into the microspaces and processes of organized criminal economies would add nuances to our understandings of such markets, which are inevitably absent from more broadly constructed accounts (Aas, 2007: 125). At the moment truly multiscalar perspectives on organized crime are difficult to construct and sustain given something of a scalar imbalance within existing literatures. More systematic and comprehensive critical syntheses of these existing and future literatures are imperatives to enhancing our understandings of many questions pertaining to the spatialities of organized crime.

A number of specific questions have emerged within this book that would, likely, repay further empirical investigation. These include the questions of the continued significance of the traditional basis for criminal organization, namely, place and ethnicity, under more post-Fordist conditions and within market contexts now characterized more strongly than was the case in the past by networked criminal forms. Furthermore, it would appear that the mechanisms of network formation and organization are relatively little understood, in anything but outline, as are the impacts of networking on the often diverse criminal organizations within these cooperative forms. While there appears to be some consensus within the literatures of organized crime around the growing hegemony of networked organization, it remains still somewhat opaque to the criminological and geographical gaze.

While network ontologies are relatively well established within the literatures of economic geography, all be they on the basis of having attained a conceptual purchase that is yet to be matched by bodies of empirical analysis, they are more emergent within the literatures of criminology generally and organized crime specifically. Translating network ontologies into programs of empirical analysis within the context of the global illicit economy of organized crime is a task crucial to sustaining and enhancing the value of these perspectives. Without this work, they are likely to have little long-term disruptive effect on our understandings of organized crime. Questions of the application of GPN perspectives (see Chapter 6; Coe et al., 2008; Hudson, 2004) to organized criminal economies and the transformative roles of criminal mobilities within the spaces through which they move emerged as most pressing from these earlier discussions.

Conceptually, there is much work to be done bringing together the literatures of organized crime and economic geography. There are debates developing around points of convergence here (Hall, 2010b, 2012, 2013; Hudson, 2014, 2016), but what emerges most strongly from these literatures, and at many points within this book, is how much is to be done here conceptually and how little empirical analysis currently exists that is designed to directly illuminate this nexus. While economic geographers have contributed much to nuanced, critical understandings of the global economy, they have done so to date with almost no reference to its illicit realms. While the conceptual tool kit that economic geography has developed is rich, varied, and dynamic, we know little at the moment of its transferability, designed, as it is, largely to speak to licit globalizations, to the illicit realms of the contemporary global economy. Without doubt there will be significant overlaps between the licit and the illicit aspects of the contemporary global economy—one might point, for example, to the recent rise of China as both a hub of licit manufacturing and also a world center for the manufacture of counterfeit goods of many kinds (Chow, 2003; Phillips, 2005). However, to suggest that illicit globalizations are merely shadows of their licit counterparts is to ignore the likelihood of significant differences in their global spatialities, the empirical conditions that underpin them, and their theoretical explanation (Zook, 2003: 1277). There are many parts of the world, for example, that seem, according to prevailing discourses of the global economy, to be largely disengaged from circuits of contemporary global commodity movement and capital circulation. However, acknowledging the illicit as an important yet overlooked aspect of the global economic landscape paints a modified picture that, in some cases, highlights apparently peripheral regions as actually dynamic centers of criminal economic activity (Berlusconi et al., 2017: 2; Nordstrom, 2011: 13). Indeed, it may be the very conditions therein that preclude articulations with the licit economy that ensure their suitability for the illicit economy (Castells, 2000: 162; Hall, 2010a: 844, 2013: 376; Zook, 2003: 1263). Therefore, we should acknowledge that the models of the contemporary global economy that have been built within economic geography and cognate disciplines have not done enough to accommodate illicit economic markets into the knowledges they have advanced. This book has argued that it is important to recognize this partiality and to consider the transferability of these models across a wider economic terrain and one that, crucially, includes illicit as well and licit dimensions. Indeed, given much recent corporate practice, the licit–illicit distinction threatens to become of less and less value in underpinning mappings of the global economy.

Finally, in parallel to all of the empirical and conceptual endeavors outlined above, are questions about the application of economic geographies of organized crime to policy and to other responses to the problems it poses. Attempts to tackle organized crime have, overwhelmingly, failed to manage, let alone curtail, its economies and its impacts on people, markets, regions, and politics. The imperative for more sophisticated and realistic policy responses to organized crime has long been recognized. This is a project to which economic geography perspectives can undoubtedly, and should seek to, make valuable, critical contributions.

REFERENCES

Aas, K. F. (2007) *Globalization and Crime*. London: SAGE.
Abadinsky, H. (2013) *Organized Crime* (10th ed.). Belmont, CA: Wadsworth, Cengage Learning.
Abdullah, A. F. L., Chang K. H., Desa, W. N. S. M., Sulaiman, M., Hamden, R., and Kunalan, V. (2014) Clandestine drug laboratory: Emergence, types, factors and problems. *Health and the Environment Journal*, 5(2), 11–27.
Abraham, I., and van Schendel, W. (2005) Introduction: The making of illicitness. In W. van Schendel and I. Abraham (Eds.), *Illicit Flows and Criminal Things: States, Borders and the Other Side of Globalization*. Bloomington: Indiana University Press.
Adler, P. (1985) *Wheeling and Dealing: An Ethnography of an Upper Level Drug Dealing and Smuggling Community*. New York: Columbia University Press.
Akee, R., Basu, A. K., Bedi, A., and Chau, N. H. (2014) Transnational trafficking, law enforcement, and victim protection: A middleman trafficker's perspective. *Journal of Law and Economics*, 57(2), 349–386.
Albanese, J. (2005) North American organised crime. In M. Galeotti (Ed.), *Global Crime Today: The Changing Face of Organised Crime*. London: Routledge.
Albini, J. (1971) *The American Mafia: Genesis of a Legend*. New York: Appleton Century Crofts.
Albini, J. L., Kutushev, V., Rogers, R. E., Moiseev, V., Shabalin, V., and Anderson, J. (1997) Russian organized crime: Its history, structure and function. In P. J. Ryan and G. E. Rush (Eds.), *Understanding Organized Crime in Global Perspective: A Reader*. London: SAGE.
Alexander, H. (2015, November 11) Colombia overtakes Peru to be world's largest producer of cocaine. Retrieved April 21, 2016, from *www.telegraph.co.uk/news/worldnews/southamerica/colombia/11989635/Colombia-overtakes-Peru-to-be-the-worlds-largest-producer-of-cocaine.html*.

Allen, C. M. (2005) *An Industrial Geography of Cocaine*. New York: Routledge.
Allum, F. (2006) *Camorristi, Politicians and Businessmen: The Transformation of Organized Crime in Post-War Naples*. Leeds, UK: Maney.
Álvarez, M. D. (2002) Illicit crops and bird conservation priorities in Colombia. *Conservation Biology*, 16(4), 1086–1096.
Amin, A. (1994) *Post-fordism: A Reader*. Oxford, UK: Blackwell.
Anderson, R., Barton, C., Böhme, R., Clayton, R., van Eeten, M. J. G., Levi, M., et al. (2013) Measuring the cost of cybercrime. In R. Böhme (Ed.), *The Economics of Information Security and Piracy*. Heidelberg, Germany: Springer.
Andreas, P. (2002) Transnational crime and economic globalization. In M. Berdal and M. Serrano (Eds.), *Transnational Organized Crime and International Security: Business as Usual?* Boulder, CO: Lynne Rienner.
Andreas, P. (2005) Criminalizing consequences of sanctions: Embargo busting and its legacy. *International Studies Quarterly*, 49, 335–360.
Andreas, P. (2014) *Smuggler Nation: How Illicit Trade Made America*. Oxford, UK: Oxford University Press.
Aning, K. (2007) Are there emerging West African criminal networks?: The case of Ghana. *Global Crime*, 8(3), 193–212.
Antonopoulos, G. A., Hobbs, D., and Hornsby, R. (2011) A soundtrack to (illegal) entrepreneurship: The counterfeit CD/DVD market in a Greek provincial city. *British Journal of Criminology*, 51(5), 804–822.
Aoyama, Y., Murphy, J. T., and Hanson, S. (2011) *Key Concepts in Economic Geography*. London: SAGE.
Arlacchi, P. (2004) *Addio Cosa Nostra*. Milan: Biblioteca Univerzale Rizzoli.
Atuesta Becerra, L. H. (2014) Addressing the costs of prohibition: Internally displaced populations in Colombia and Mexico. In LSE Expert Group on the Economics of Drug Policy (Ed.), *Ending the Drug Wars: Report of the LSE Expert Group on the Economics of Drug Policy*. London: London School of Economics.
Bagley, B. (2005) Globalisation and Latin American and Caribbean organised crime. In M. Galeotti (Ed.), *Global Crime Today: The Changing Face of Organised Crime*. London: Routledge.
Bair, J., and Gereffi, G. (2001) Local clusters in global chains: The causes and consequences of export dynamism in Torreon's blue jeans industry. *World Development*, 29(11), 1885–1903.
Baker, P. (1993) The role of the small firm in locality restructuring. *Area*, 25(1), 37–44.
Bayart, J. F., Ellis, S., and Hibou, B. (1999) Introduction. In J. F. Bayart, S. Ellis, and B. Hibou (Eds.), *The Criminalization of the State in Africa*. Oxford, UK: Currey.
BBC World Service. (2007, November) *Africa's Cocaine Coast* (Radio documentary).
Bell, D. (1953) Crime as an American way of life. *Antioch Review*, 13(2), 131–154.
Berlusconi, G., Aziani, A., and Giommoni, L. (2017) The determinants of heroin

flows in Europe: A latent space approach. *Social Networks*. Retrieved from *https://doi.org/10.1016/j.socnet.2017.03.012*.
Bhattacharyya, G. (2005) *Traffick: The Illicit Movement of People and Things*. London: Pluto Press.
Birtchnell, T., Savitzky, S., and Urry, J. (Eds.). (2015) *Cargomobilities: Moving Materials in a Global Age*. Abingdon, UK: Routledge.
Block, A., and Chambliss, W. (1981) *Organizing Crime*. New York: Elsevier.
Blok, A. (1974) *The Mafia of a Sicilian Village, 1860–1960*. New York: Harper.
Boivin, R. (2013) Drug trafficking networks in the world economy. In C. Morselli (Ed.), *Crime and Networks*. New York: Routledge.
Boivin, R. (2014a) Macrosocial network analysis: The case of transnational drug trafficking. In A. J. Masys (Ed.), *Networks and Network Analysis for Defence and Security, Lecture Notes in Social Networks*. New York: Springer.
Boivin, R. (2014b) Risks, prices and positions: A social network analysis of illegal drug trafficking in the world economy. *International Journal of Drug Policy, 25*(2), 235–243.
Boivin, R. (n.d.) *Drug trafficking networks in the world-economy*. Unpublished paper, retrieved October 25, 2013, https://www.researchgate.net/publication/265069568_Drug_trafficking_networks_in_the_world-economy.
Bosco, F. J. (2006) Actor–network theory, networks and relational approaches in human geography. In S. Aitken and G. Valentine (Eds.), *Approaches to Human Geography* London: SAGE.
Bourgois, P. (1995) *In Search of Respect*. Cambridge, UK: Cambridge University Press.
Bowden, M. (2012) *Killing Pablo: The Hunt for the Richest, Most Powerful Criminal in History*. London: Atlantic Books.
Boyce, G. A., Banister, J. M., and Slack, J. (2015) You and what army?: The state, and Mexico's war on drugs. *Territory, Politics, Governance, 3*(4), 446–468.
Brooks, A. (2012) Networks of power and corruption: The trade of Japanese used cars to Mozambique. *Geographical Journal, 178*(1), 80–92.
Brooks, A. (2014) Controversial, corrupt and illegal: Ethical implications of investigating difficult topics. Reflections on fieldwork in southern Africa. In J. Lunn (Ed.), *Fieldwork in the Global South: Ethical Challenges and Dilemmas*, Abingdon, UK: Routledge.
Brooks, A. (2015) *Clothing Poverty: The Hidden World of Fast Fashion and Second-Hand Clothes*. London: Zed Books.
Brophy, S. (2008) Mexico: Cartels, corruption and cocaine: A profile. *Global Crime, 9*(3), 248–261
Brown, E., and Cloke, J. (2004) Neoliberal reform, governance and corruption in the South: Assessing the international anti-corruption crusade. *Antipode, 36*(2), 272–294.
Brown, E., and Cloke, J. (2007) Shadow Europe: Alternative European financial geographies. *Growth and Change, 38*(2), 304–327.
Browning, F., and Gerassi, J. (1980) *The American Way of Crime*. New York: Putnam.

Bryman, A. (2012) *Social Research Methods*. Oxford, UK: Oxford University Press.
Bucquoye, A., Verpoest, K., Defruytier, M., & Vander Beken, T. (2005) European road transport of goods. In T. Vander Beken (Ed.), *Organised Crime and Vulnerability of Economic Sectors: The European Transport and Music Sector*. Antwerp-Apeldoorn, The Netherlands: Maklu.
Bunker, R. J. (2011a, November) The growing Mexican cartel and vigilante war in cyberspace: Information offensives and counter-offensives. Retrieved from *Small Wars Journal*, June 28, 2016, *http://smallwarsjournal.com/jrnl/art/the-growing-mexican-cartel-and-vigilante-war-in-cyberspace*.
Bunker, R. J. (Ed.). (2011b) *Narcos over the Border: Gangs, Cartels and Mercenaries*. Abingdon, UK: Routledge.
Calderoni, F. (2011) Where is the mafia in Italy?: Measuring the presence of the mafia across Italian provinces. *Global Crime, 12*(1), 41–69.
Cameron, A. (2008) Crisis? What crisis?: Displacing the spatial imaginary of the fiscal state. *Geoforum, 39*(3), 1145–1154.
Campana, P. (2011) Eavesdropping on the mob: The functional diversification of mafia activities across Italian provinces. *European Journal of Criminology, 8*(3), 213–228.
Campbell, D. (2016, January 23) One last job: The inside story of the Hatton Garden heist. *The Guardian*. Retrieved April 21, 2016, from *www.theguardian.com/uk-news/2016/jan/23/one-last-job-inside-story-of-the-hatton-garden-heist*.
Carter, D. L. (1997) International organized crime: Emerging trends in entrepreneurial culture. In P. J. Ryan and G. E. Rush (Eds.), *Understanding Organized Crime in Global Perspective: A Reader*. London: SAGE.
Castells, M. (2000) *End of Millennium*. Oxford, UK: Blackwell.
Catanzaro, R. (1992) *Men of Respect*. New York: Free Press.
Cauduro, A., Di Nicola, A., Fonio, C., Nuvoloni, A. A., and Ruspini, P. (2009) Innocent when you dream: Clients and trafficked women in Italy. In A. Di Nicola, A. Cauduro, M. Lombardi, and P. Ruspini (Eds.), *Prostitution and Trafficking: Focus on Clients*. New York: Springer.
Caulkins, J. P. (2014) Effects of prohibition, enforcement and interdiction on drug use. In LSE Expert Group on the Economics of Drug Policy (Ed.), *Ending the Drug Wars: Report of the LSE Expert Group on the Economics of Drug Policy*. London: London School of Economics.
Chambliss, W. (1978) *On the Take*. Bloomington: Indiana University Press.
Chandra, S., and Joba, J. (2015) Transnational cocaine and heroin flow networks in Western Europe: A comparison. *International Journal of Drug Policy, 26*(8), 772–780.
Chandra, S., Yu, Y.-L., and Bihani, V. (2016) How MDMA flows across the USA: Evidence from price data. *Global Crime, 18*(2), 122–139.
Channel 4 [UK]. (2010, August–September) *The Hunt for Britain's Sex Traffickers* [3 episodes].
Chibnall, S. (2009) Travels in ladland: The British gangster film cycle, 1998–2001. In R. Murphy (Ed.), *The British Cinema Book*. London: Palgrave.
Chiodelli, F., Hall, T., and Hudson, R. (Eds.). (2017) *The Illicit and Illegal*

in Regional and Urban Governance and Development: Corrupt Places. Abingdon, UK: Routledge.

Chiodelli, F., Hall, T., Hudson, R., and Moroni, S. (2017) The grey governance and development of cities and regions: The variable relationship between (il)legal and (il)licit. In F. Chiodelli, T. Hall, and R. Hudson (Eds.), *The Illicit and Illegal in Regional and Urban Governance and Development: Corrupt Places*. Abingdon, UK: Routledge.

Chouvy, P.-A. (2005a) Morocco said to produce nearly half of world's hashish supply. *Jane's Intelligence Review, 17*(11), 32–35.

Chouvy, P.-A. (2005b) Morocco's smuggling rackets: Hashish, people and contraband. *Jane's Intelligence Review, 17*(12), 40–43.

Chouvy, P.-A. (2009) *Opium: Uncovering the Politics of the Poppy*. London: I. B. Taurus.

Chouvy, P.-A. (2013) *An Atlas of Trafficking in Southeast Asia: The Illegal Trade in Arms, Drugs, People, Counterfeit Goods and Natural Resources in Mainland Southeast Asia*. London: I. B. Taurus.

Chow, D. C. K. (2003) Organized crime, local protectionism, and the trade in counterfeit goods in China. *China Economic Review, 14*(4), 473–484.

Christophers, B. (2011) Follow the thing: Money. *Environment and Planning D: Society and Space, 29*(6), 1068–1084.

Chung, J., Daneshkhu, S., and Masters, B. (2009, June 24) A bitter dividend. *Financial Times* [UK].

CIBJO: The World Jewellery Confederation. (2007) *Believe in Me: A Jewellery Retailer's Guide to Consumer Trust*. London: Author.

Clemens, J. (2008) Opium in Afghanistan: Prospects for the success of source country drug control policies. *Journal of Law and Economics, 51*(3), 407–432.

Clifford, N., French, S., and Valentine, G. (2010) *Key Methods in Geography*. London: SAGE.

Cobham, A., Janský, P., and Meinzer, M. (2015) The financial secrecy index: Shedding new light on the geography of secrecy. *Economic Geography, 91*(3), 237–391.

Coe, N. M., Dicken, P., & Hess, M. (2008) Global production networks: Realizing the potential. *Journal of Economic Geography, 8*(3), 271–295.

Coe, N. M., Kelly, P., and Yeung, H. (2007) *Economic Geography: A Contemporary Introduction*. Oxford, UK: Blackwell.

Cohen, P. T. (2009) The post-opium scenario and rubber in northern Laos: Alternative Western and Chinese models of development. *International Development of Drug Policy, 20*(5), 424–430.

Collins, J. (2014a) The economics of a new global strategy. In LSE Expert Group on the Economics of Drug Policy (Ed.), *Ending the Drug Wars: Report of the LSE Expert Group on the Economics of Drug Policy*. London: London School of Economics.

Collins, J. (2014b) Executive summary. In LSE Expert Group on the Economics of Drug Policy (Ed.), *Ending the Drug Wars: Report of the LSE Expert Group on the Economics of Drug Policy*. London: London School of Economics.

Cook, I. (2004) Follow the thing: Papaya. *Antipode, 36*(4), 642–664.
Cook, I., and Harrison, M. (2007) Follow the thing: West Indian hot pepper sauce. *Space and Culture, 10*(1), 40–63.
Corva, D. (2008) Neoliberal globalization and the war on drugs: Transnationalizing illiberal governance in the Americas. *Political Geography, 27*(2), 176–193.
Costa, T. G., and Schulmeister, G. H. (2007) The puzzle of the Iguazu tri-border area: Many questions and few answers regarding organised crime and terrorism links. *Global Crime, 8*(1), 26–39.
Crampton, J. W. (2001) Maps as social constructions: Power, communication and visualization. *Progress in Human Geography, 25*(2), 235–252.
Crampton, J. W., and Krygier, J. (2005) An introduction to critical cartography. *ACME: An International E-Journal for Critical Geographies, 4*(1), 11–33.
Crang, M. (1999) *Cultural Geography*. London: Routledge.
Cressey, D. R. (1969) *Theft of the Nation: The Structure and Operations of Organized Crime in America*. New York: Harper & Row.
Cressey, D. R. (1997) The functions and structures of criminal syndicates. In P. J. Ryan and G. E. Rush (Eds.), *Understanding Organized Crime in Global Perspective: A Reader*. London: SAGE.
Cribb, R. (2009) Introduction: Parapolitics, shadow governance and criminal sovereignty. In E. Wilson (Ed.), *Government of the Shadows: Parapolitics and Criminal Sovereignty*. London: Pluto Press.
Critchley, D. (2009) *The Origins of Organized Crime in America: The New York City Mafia, 1891–1931*. New York: Routledge.
Curtis, R., and Wendel, T. (2000) Toward the development of a typology of illegal drug markets. *Crime Prevention Studies, 11*, 121–152.
Cusicanqui, S. R. (2005) "Here, even legislators chew them": Coca leaves and identity politics in northern Argentina. In W. von Schendel and I. Abraham (Eds.), *Illicit Flows and Criminal Things: States, Borders and the Other Side of Globalization*. Bloomington: Indiana University Press.
Daniels, P. W. (2004) Urban challenges: The formal and informal economies in mega-cities. *Cities, 21*(6), 501–511.
Daudelin, J. (2010) *Moving Frontiers: Patterns of Drug Violence in the Americas*. Ottawa, ON, Canada: Norman Paterson Schools of International Affairs, Carleton University.
Davidson, R. L. (1997) Asian gangs and Asian organized crime in Chicago. In P. J. Ryan and G. E. Rush (Eds.), *Understanding Organized Crime in Global Perspective: A Reader*. London: SAGE.
Davis, M. (2007) *Planet of Slums*. London: Verso.
Decker, S. H., and Townsend Chapman, M. (2008) *Drug Smugglers on Drug Smuggling*. Philadelphia: Temple University Press.
Deneault, A. (2007) Tax havens and criminology. *Global Crime, 8*(3), 260–270.
Depropris, L., and Wei, P. (2007) Governance and competitiveness in the Birmingham Jewellery District. *Urban Studies, 44*(12), 2465–2486.
Deverteuil, G., and Wilton, R. (2009) The geographies of intoxicants: From production and consumption to regulation, treatment and prevention. *Geography Compass, 3*(1), 478–494.

Dick, H. (2009) The shadow economy: Markets, crime and the state. In E. Wilson (Ed.), *Government of the Shadows: Parapolitics and Criminal Sovereignty*. London: Pluto Press.

Dicken, P. (2011) *Global Shift: Mapping the Changing Contours of the World Economy* (6th ed.). New York: Guilford Press.

Dicken, P., Kelly, P. F., Olds, K., and Yeung, H. W.-C. (2001) Chains and networks, territories and scales: Towards a relational framework for analysing the global economy. *Global Networks, 1*(2), 89–112.

Dickson-Gilmore, J., and Woodiwiss, M. (2008) The history of Native Americans and the misdirected study of organized crime. *Global Crime, 9*(1–2), 66–83.

Dittmer, J. (2007) The tyranny of the serial: Popular geopolitics, the nation and comic book discourse. *Antipode, 39*(2), 247–268.

Dittmer, J. (2010) *Popular Culture, Geopolitics and Identity*. Lanham, MD: Rowman & Littlefield.

Dodds, K. (2005) Screening geopolitics: James Bond and the early Cold War films (1962–1967). *Geopolitics, 10*(2), 266–289.

Dodds, K. (2013) "I'm still not crossing that": Borders, dispossession, and sovereignty in *Frozen River* (2008). *Geopolitics, 18*(3), 560–583.

Dorling, D., and Lee, C. (2016) *Geography*. London: Profile Books.

Dorn, N., Murji, K., and South, N. (1992) *Traffickers*. London: Routledge.

Dourojeanni, M. (1992) Environmental impact of coca cultivation and cocaine production in the Amazon region of Peru. *UN-ODCCP Bulletin on Narcotics, 2*, 37–53.

Dunlap, E., Johnson, B., and Manwar, A. (1994) A successful female crack dealer: Case study of a deviant career. *Deviant Behaviour, 15*(1), 1–25.

Dunn, G. (1997) Major mafia gangs in Russia. In P. Williams (Ed.), *Russian Organised Crime: The New Threat?* London: Frank Cass.

Duthie, E. (2014) *Celebrities in Conservation: Are They a Hindrance or a Help?* Unpublished MSc thesis, Imperial College, London. Retrieved from www.iccs.org.uk/wp-content/uploads/2011/10/Duthie_Elizabeth_ConSci_2014.pdf.

Ebbe, O. N. I. (1999) Political–criminal nexus: The Nigerian case: Slicing Nigeria's "national cake." *Trends in Organized Crime, 4*(3), 29–59.

The Economist. (2013a, February 16) Enduring charms. *The Economist* (Special Report: Offshore Finance).

The Economist. (2013b, February 16) Not a palm tree in sight. *The Economist* (Special Report: Offshore Finance).

The Economist. (2013c, February 16) Storm survivors. *The Economist* (Special Report: Offshore Finance).

Edberg, M. C. (2001) Drug traffickers as social bandits: Culture and drug trafficking in northern Mexico and the border region. *Journal of Contemporary Criminal Justice, 17*(3), 259–277.

Edelman, M. (1988) *Constructing the Political Spectacle*. Chicago: University of Chicago Press.

Egmond, F. (1993) *Underworlds: Organised Crime in the Netherlands, 1650–1800*. Cambridge, UK: Polity Press.

Eiss, P. K. (2014) The narcomedia: A reader's guide. *Latin American Perspectives*, 41(2), 78–98.
Ellis, S. (2009) West Africa's international drug trade. *African Affairs*, 108(431), 171–196.
Escobar, R. (2010) *Escobar: Drugs. Guns. Money. Power.* London: Hodder.
Ettinger, N., and Bosco, F. (2004) Thinking through networks and their spatiality: A critique of the US (public) war on terrorism and its geographic discourse. *Antipode*, 36(2), 249–271.
European Union. (2015) Trafficking in human beings. Retrieved May 5, 2017, from *https://ec.europa.eu/home-affairs/what-we-do/policies/organized-crime-and-human-trafficking/trafficking-in-human-beings_en*.
Europol. (2007) *OCTA—EU Organised Crime Threat Assessment*. The Hague, The Netherlands: Author.
Europol. (2013) *SOCTA—EU Serious and Organised Crime Threat Assessment*. The Hague, The Netherlands: Author.
Farrell, G. (1998) A global empirical review of drug crop eradication and United Nations' crop substitution and alternative development strategies. *Journal of Drug Issues*, 28(2), 395–436.
Farrell, G., and Thorne, J. (2005) Where have all the flowers gone?: Evaluation of the Taliban crackdown against opium poppy cultivation in Afghanistan. *International Journal of Drug Policy*, 16(2), 81–91.
Farwell, J. R. (2014) The media strategy of ISIS. *Survival: Global Politics and Strategy*, 56(6), 49–55.
Felbab-Brown, V. (2014) Improving supply-side policies: Smarter eradication, interdiction and alternative livelihoods—and the possibility of licensing. In LSE Expert Group on the Economics of Drug Policy (Ed.), *Ending the Drug Wars: Report of the LSE Expert Group on the Economics of Drug Policy*. London: London School of Economics.
Ferragut, S. (2012) *Organized Crime, Illicit Drugs and Money Laundering: The United States and Mexico*. London: Chatham House.
Fickley, A. (2011) "The focus has to be on helping people make a living": Exploring diverse economies and alternative economic spaces. *Geography Compass*, 5(5), 237–248.
Financial Action Task Force. (2011) *Money Laundering Risks Arising from Trafficking in Human Beings and Smuggling of Migrants*. Paris: Author.
Finckenauer, J. O. (2005) Problems of definition: What is organized crime? *Trends in Organized Crime*, 8(3), 63–83.
Firestone, T. A. (1993) Mafia memoirs: What they can tell us about organized crime. *Journal of Contemporary Criminal Justice*, 9(3), 197–220.
Fisher, B., Ialomiteanu, A. R., Russell, C., Rehm, J., and Mann, R. E. (2016) Public opinion towards cannabis control in Ontario: Strong but diversified support for reforming control of both use and supply. *Canadian Journal of Criminology and Criminal Justice*. Retrieved from *http://dx.doi.org/10.3138/CJCCJ.2015E.43*.
Fisher, R. (2013) "A gentleman's handshake": The role of social capital and trust in transforming information into usable knowledge. *Journal of Rural Studies*, 31(1), 13–22.

Fleetwood, J. (2014) *Drug Mules: Women in the International Cocaine Trade.* Basingstoke, UK: Palgrave Macmillan.

Foltz, J., Jackson, J., Oberg, A., and The European Futures Observatory. (2008) *The Globalisation of Crime.* Ipswich, UK: The European Futures Observatory.

Foster, P. (2017, July 12) EU anti-migrant naval mission in Med is a "failure," Lords report finds. *The Telegraph* [UK]. Retrieved July 12, 2017, from *www.telegraph.co.uk/news/2017/07/11/eu-anti-migrant-naval-mission-med-failure-lords-report-finds.*

Frankfort-Nachmias, C., Nachmias, D., and DeWaard, J. (2014) *Research Methods in the Social Sciences.* London: Palgrave Macmillan.

Fraser, F., and Morton, J. (1995) *Mad Frank: Memoirs of a Life of Crime.* London: Warner.

Frisby, T. (1998) The rise of organised crime in Russia: Its roots and social significance. *Europe–Asia Studies, 50*(1), 27–49.

Galeotti, M. (2005a) Introduction: Global crime today. In M. Galeotti (Ed.), *Global Crime Today: The Changing Face of Organised Crime.* London: Routledge.

Galeotti, M. (2005b) The Russian "Mafiya": Consolidation and globalisation. In M. Galeotti (Ed.), *Global Crime Today: The Changing Face of Organised Crime.* London: Routledge.

Galeotti, M. (2008) Criminal histories: An introduction. *Global Crime, 9*(1–2), 1–7.

Gamba, P., and Herold, M. (Eds.). (2009) *Global Mapping of Human Settlement: Experiences, Datasets and Prospects.* Boca Raton, FL: CRC Press.

Gambetta, D. (1993) *The Sicilian Mafia: The Business of Private Protection.* Cambridge, MA: Harvard University Press.

Garrett, B. (2013) *Explore Everything: Place-Hacking the City.* London: Verso.

Garrett, B. (2014, June 11) Access denied. *Times Higher Education* [UK].

George, R. (2015, July 2) ZeroZeroZero by Roberto Saviano, book review: The terrifying violence of the cocaine trade. *The Independent* [UK]. Retrieved June 30, 2016, from *www.independent.co.uk/arts-entertainment/books/reviews/zerozerozero-by-roberto-saviano-book-review-the-terrifying-violence-of-the-cocaine-trade-10360790.html.*

Gibler, J. (2011) *To Die in Mexico: Dispatches from Inside the Drug War.* San Francisco: City Lights Books.

Gibson-Graham, J. K. (2008) Diverse economies: Performative practices for "other worlds." *Progress in Human Geography, 32*(5), 613–632.

Gill, P. (2006) Organized crime. In E. McLaughlin and J. Muncie (Eds.), *The SAGE Dictionary of Criminology* (2nd ed.). London: SAGE.

Glenny, M. (2008) *McMafia: Crime Without Frontiers.* London: Bodley Head.

Glenny, M. (2011) *Dark Market: Cyberthieves, Cybercops and You.* London: Bodley Head.

Goodhand, J. (2009) *Bandits, borderlands and opium wars: Afghan state-building viewed from the margins.* (Markets for Peace? Working Paper No. 26). Copenhagen: DIIS.

Goodhand, J. (2011) Corrupting or consolidating the peace: The drugs economy and post-conflict peacebuilidng in Afghanistan. In D. Zaum and C. Cheng (Eds.), *Corruption and Post Conflict Peacebuilding: Selling the Peace?* London: Routledge.

Goodhand, J. (2012) Bandits, borderlands and opium wars: Afghan statebuilding viewed from the margin. In T. Wilson and H. Donnan (Eds.), *A Companion to Border Studies*. London: Wiley-Blackwell.

Goodhand, J., and Mansfield, D. (2013) Drugs and (dis)order: The opium economy, political settlements and statebuilding in Afghanistan. In C. Schetter (Ed.), *Local Politics in Afghanistan: A Century of Intervention in the Social Order*. London: Hurst.

Goodhand, J., and Sedra, M. (2015) *The Afghan Conundrum: Intervention, Statebuilding and Resistance*. London: Routledge.

Grant, R., and Oteng-Ababio, M. (2012) Mapping the invisible and real "African" economy: Urban e-waste circuitry. *Urban Geography*, 33(1), 1–21.

Grascia, A. M. (2004) Gang violence: Mara Salvatrucha—"Forever Salvador." *Journal of Gang Research*, 11(2), 29–36.

Gregory, D. (2011) The everywhere war. *Geographical Journal*, 177(3), 238–250.

Gregson, N., and Crang, M. (2017) Illicit economies: Customary illegality, moral economies and circulation. *Transactions of the Institute of British Geographers*, 42(2), 206–219.

Hall, T. (2010a) Economic geography and organised crime: A critical review. *Geoforum*, 41(6), 841–845.

Hall, T. (2010b) Where the money is: The geographies of organised crime. *Geography*, 95(1), 4–13.

Hall, T. (2012) The geography of transnational organised crime: Spaces, networks and flows. In F. Allam and S. Gilmor (Eds.), *The Routledge Handbook of Transnational Organized Crime*. Abingdon, UK: Routledge.

Hall, T. (2013) Geographies of the illicit: Globalisation and organised crime. *Progress in Human Geography*, 37(3), 366–385.

Hall, T. (2016) Biographies of illicit super-wealth. In I. Hay and J. V. Beaverstock (Eds.), *Handbook on Wealth and the Super-Rich*. Cheltenham, UK: Edward Elgar.

Hall, T. (2017, July) *Against the flow: Unpacking official discourses of illicit mobility*. Paper presented to the European Consortium for Political Research, 2nd General Conference on Organised Crime, University of Bath, UK, 7–8 July 2017.

Haller, M. (1971) Organized crime in urban society: Chicago in the twentieth century. *Journal of Social History*, 5(2), 210–234.

Hallsworth, S. (2013) *The Gang and Beyond: Interpreting Violent Street Worlds*. Basingstoke, UK: Palgrave Macmillan.

Hampton, M., and Levi, M. (1999) Fast spinning into oblivion?: Recent developments in money-laundering policies and offshore finance centres. *Third World Quarterly*, 20(3), 645–656.

Harbi, S., & Grolleau, G. (2008) Profiting from being pirated by pirating the pirates. *Kyklos*, 61, 385–390.

Harley, J. B. (1988) Maps, knowledge, and power. In D. Cosgrove and S.

Daniels (Eds.), *The Iconography of Landscape: Essays on the Symbolic Representation, Design and Use of Past Environments*. Cambridge, UK: Cambridge University Press.

Harvey, D. (1989) *The Condition of Postmodernity: An Enquiry into the Origins of Cultural Change*. Oxford, UK: Blackwell.

Hastings, J. V. (2009) Geographies of state failure and sophistication in maritime piracy hijackings. *Political Geography*, 28(4), 213–223.

Hastings, J. V. (2015) The economic geography of North Korean drug trafficking networks. *Review of International Political Economy*, 22(1), 162–193.

Haufler, V. (2010) The Kimberley Process Certification Scheme: An innovation in global governance in conflict prevention. *Journal of Business Ethics*, 89(Suppl. 4), 403–416.

Hay, C. (1996) Narrating crisis: The discursive construction of the "winter of discontent." *Sociology*, 30(2), 253–277.

Haynes, D. (2017, July 12) Royal Navy mission fails to curb flow of migrants across Mediterranean. *The Times* [UK].

Healey, M., and Ilbery, B. (1990) *Location and Change: Perspectives on Economic Geography*. Oxford, UK: Oxford University Press.

Hepple, L. W. (1992) Metaphor, geopolitical discourse and the military in South America. In T. J. Barnes and J. S. Duncan (Eds.), *Writing Worlds: Discourse, Text and Metaphor in the Representation of Landscape*. London: Routledge.

Hernàndez, A. (2010) *Los Señores del Narco*, Barcelona, Spain: Grijalbo.

Hernàndez, A. (2013a) *México en Llamas: El Legado de Calderon*. Barcelona, Spain: Grijalbo.

Hernàndez, A. (2013b) *Narcoland: The Mexican Drug Lords and Their Godfathers*. London: Verso.

Herwartz, H., Tafenau, E., and Schneider, F. (2015) One share fits all?: Regional variations in the extent of the shadow economy in Europe. *Regional Studies*, 49(9), 1575–1587.

Hess, H. (1973) *Mafia and Mafiosi: The Structure of Power*. Lexington, MA: D. C. Heath.

Hesse, B. (2011) Introduction: The myth of "Somalia." In B. Hesse (Ed.), *Somalia: State Collapse, Terrorism and Piracy*. Abingdon, UK: Routledge.

Hetzer, H., and Walsh, J. (2014) Pioneering cannabis regulation in Uruguay. *NACLA Report on the Americas*, 47(2), 33–35.

Hignett, K. (2005) Organised crime in East Central Europe: The Czech Republic, Hungary and Poland. In M. Galeotti (Ed.), *Global Crime Today: The Changing Face of Organised Crime*. London: Routledge.

Hill, P. (2005) The changing face of the Yakuza. In M. Galeotti (Ed.), *Global Crime Today: The Changing Face of Organised Crime*. London: Routledge.

Hobbs, D. (1988) *Doing the Business*. Oxford, UK: Oxford University Press.

Hobbs, D. (1995) *Bad Business: Professional Crime in Modern Britain*. Oxford, UK: Oxford University Press.

Hobbs, D. (1998) The case against: There is not a global crime problem. *International Journal of Risk, Security, and Crime Prevention*, 3, 139–146.

Hobbs, D. (2001) The firm: Organizational logic and criminal culture on a shifting terrain. *British Journal of Criminology, 41*(4), 549–560.

Hobbs, D. (2004) The nature and representation of organised crime in the United Kingdom. In C. Fijnaut and L. Paoli, L. (Eds.), *Organised Crime in Europe: Concepts, Patterns and Control Policies in the European Union and Beyond*. Dordrecht, The Netherlands: Springer.

Hobbs, D. (2013) *Lush Life: Constructing Organized Crime in the UK*. Oxford, UK: Oxford University Press.

Hobbs, D., and Antonopoulos, G. A. (2013) "Endemic to the species": Ordering the "other" via organised crime. *Global Crime, 14*(1), 27–51.

Hobbs, D., and Antonopoulos, G. A. (2014) How to research organised crime. In L. Paoli (Ed.), *The Oxford Handbook of Organized Crime*. Oxford, UK: Oxford University Press.

Hobbs, D., and Dunnighan, C. (1998) Global organised crime: Context and pretext. In V. Ruggiero, N. South, and I. Taylor (Eds.), *The New European Criminology: Crime and Social Order in Europe*. London: Routledge.

Hobsbawm, E. (1972) *Bandits*. London: Pelican.

Hoggart, K., Davies, A., and Lees, L. (2002) *Researching Human Geography*. London: Arnold.

Holmes, L. (2016) *Advanced Introduction to Organised Crime*. Cheltenham, UK: Edward Elgar.

Hornsby, R., and Hobbs, D. (2007) A zone of ambiguity: The political economy of cigarette bootlegging. *British Journal of Criminology, 47*(4), 551–571.

Hudson, A. (2000) Offshoreness, globalization and sovereignty: A postmodern geo-political economy? *Transactions of the Institute of British Geographers, 25*(3), 269–283.

Hudson, R. (2004) Conceptualizing economies and their geographies: Spaces, flows and circuits. *Progress in Human Geography, 28*, 447–471.

Hudson, R. (2005) *Economic Geographies: Circuits, Flows and Spaces*. London: SAGE.

Hudson, R. (2014) Thinking through the relationships between legal and illegal activities and economies: Spaces, flows and pathways. *Journal of Economic Geography, 14*(4), 775–795.

Hudson, R. (2016) *Approaches to Economic Geography: Towards a Geographical Political Economy*. Abingdon, UK: Routledge.

Hughes, C. E., and Stevens, A. (2010) What can we learn from the Portuguese decriminalization of drugs? *British Journal of Criminology, 50*(6), 999–1022.

Ianni, F., and Ianni, E. (1972) *A Family Business: Kinship and Social Control in Organized Crime*. New York: Russell Sage Foundation.

Inkster, N., and Comolli, V. (2012) *Drugs, Insecurity and Failed States: The Problems of Prohibition*. Abingdon, UK: Routeldge/IISS.

International Labour Organization. (2012) *Global Estimate of Forced Labour 2012: Results and Methodology*. Geneva, Switzerland: Author.

Interpol. (2014) *Against Organized Crime: Interpol Trafficking and Counterfeiting Casebook. 2014*. Lyon, France: Author.

Jakobi, A. P. (2013) *Common Goods and Evils?: The Formation of Global Crime Governance*. Oxford, UK: Oxford University Press.

Jamieson, A. (1995) The transnational dimension of Italian organized crime. *Transnational Organized Crime, 1–2*, 151–172.

Jenks, C., and Lorentzen, J. J. (1997) The Kray fascination. *Theory, Culture and Society, 14*(3), 87–107.

Jesperson, S. (2014) Bridging the research and policy divide on organised crime. *European Review of Organised Crime, 1*(2), 147–157.

Jesperson, S. (2017) Filling governance and development vacuums: A role for development actors or criminal groups? In F. Chiodelli, T. Hall, and R. Hudson (Eds.), *The Illicit and Illegal in the Governance and Development of Cities and Regions: Corrupt Places*. Abingdon, UK: Routledge.

Jones, R. (2014) *Border Wars*: Narratives and images of the US–Mexico border on TV. *ACME: An International E-Journal for Critical Geographies, 13*(3), 530–550.

Kahane, L. H. (2013) Understanding the interstate export of crime guns: A gravity model approach. *Contemporary Economic Policy, 31*(3), 618–634.

Kar, D., & Spanjers, J. (2014) *Illicit Financial Flows from Developing Countries, 2002–2012*. Washington, DC: Global Financial Integrity.

Karlsson, C., Andersson, M., and Norman, T. (Eds.). (2015) *Handbook of Research Methods and Applications in Economic Geography*. Cheltenham, UK: Edward Elgar.

Kelly, P. (2006) A critique of critical geopolitics. *Geopolitics, 11*(1), 24–53.

Kenney, M. (2007) The architecture of drug trafficking: Network forms of organisation in the Columbian cocaine trade. *Global Crime, 8*(3), 233–259.

Kiely, R. (1998) Globalization, post-fordism and the contemporary context of development. *International Sociology, 13*(1), 95–115.

Kilcullen, D. (2013) *Out of the Mountains: The Coming Age of the Urban Guerrilla*. Oxford, UK: Oxford University Press.

Kilma, N. (2011) The goods transport network's vulnerability to crime: Opportunities and control weaknesses. *European Journal on Criminal Policy and Research, 17*(3), 203–219.

Kirby, S., and Penna, S. (2011) Policing mobile criminality: Implications for police forces in the UK. *Policing: An International Journal of Police Strategies and Management, 34*(2), 182–197.

Klausen, J. (2015) Tweeting the *Jihad*: Social media networks of Western foreign fighters in Syria and Iraq. *Studies in Conflict and Terrorism, 38*(1), 1–22.

Kleemans, E. R. (2008) Introduction to special edition: Organized crime, terrorism and European criminology. *European Journal of Criminology, 5*(1), 5–12.

Kleemans, E. R., and van de Bunt, H. G. (1999) The social embeddedness of organized crime. *Transnational Organized Crime, 5*(1), 19–36.

Knepper, P. (2009) *The Invention of International Crime: A Global Issue in the Making, 1881–1914*. London: Palgrave Macmillan.

Knepper, P. (2011) *International Crime in the 20th Century: The League of Nations Era, 1919–1939*. London: Palgrave Macmillan.

Krasner, S. D., and Risse, T. (2014) External actors, state-building, and service provision in areas of limited statehood: Introduction. *Governance: An International Journal of Policy, Administration and Institutions*, 27(4), 545–567.

Kupatadze, A. (2007) Radiological smuggling and uncontrolled territories: The case of Georgia. *Global Crime*, 8(1), 40–57.

Lajous, A. M. (2014) The constitutional costs of the "war on drugs." In LSE Expert Group on the Economics of Drug Policy (Ed.), *Ending the Drug Wars: Report of the LSE Expert Group on the Economics of Drug Policy*. London: London School of Economics.

Lane, C., and Probert, J. (2006) Domestic capabilities and global production networks in the clothing industry: A comparison of German and UK firms' strategies. *Socio-Economic Review*, 4(1), 35–67.

Larke, G. S. (2003) Organized crime: Mafia myths in films and television. In P. Mason (Ed.), *Criminal Visions: Media Representations of Crime and Justice*. Collumpton, UK: Wilan.

Lash, S. M., and Urry, J. (1987) *The End of Organized Capitalism*. Madison: University of Wisconsin Press.

Lash, S. M., and Urry, J. (1993) *Economies of Signs and Space*. London: SAGE.

Lavezzi, A. M. (2008) Economic structure and vulnerability to organised crime: Evidence from Sicily. *Global Crime*, 9(3), 198–220.

Laville, S. (2008, November 19) Cocaine users are destroying the rainforest—at 4 square metres per gram. *The Guardian* [UK]. Retrieved May 9, 2017, from www.theguardian.com/world/2008/nov/19/cocaine-rainforests-columbia-santos-calderon.

Law, J., and Urry, J. (2004) Enacting the social. *Economy and Society*, 33(3), 390–410.

Lee, R. (2008) The Triborder–terrorism nexus. *Global Crime*, 9(4), 332–347.

Legge, J. (2013, September 18) Attacks on her family, headless animals being sent to her home and several death threats: A sure sign Anabel Hernandez is the woman the Mexican drug barons fear. *The Independent* [UK]. Retrieved from www.independent.co.uk/news/world/americas/attacks-on-her-family-headless-animals-being-sent-to-her-home-and-several-death-threats-a-sure-sign-8824947.html.

Lemahieu, H., Sampaio, A., and Comolli, V. (Eds.). (2015) *The Strategic Implications of Organised Criminal Markets. Conference Report*. London: IISS.

Lepawsky, J., and McNabb, C. (2010) Mapping international flows of electronic waste. *Canadian Geographer*, 54(2), 177–195.

Levi, M. (2002) The organization of serious crimes. In M. Maguire, R. Morgan, and R. Reiner (Eds.), *The Oxford Handbook of Criminology*. Oxford, UK: Oxford University Press.

Levi, M. (2008) *The Phantom Capitalists: The Organization and Control of Long Firm Fraud*. Abingdon, UK: Routledge.

Levi, M. (2009) Suite revenge?: The shaping of folk devils and moral panics about white-collar crimes. *British Journal of Criminology*, 49(1), 48–67.

Levi, M. (2014) Thinking about organised crime: Structure and threat. *The RUSI Journal*, 159(1), 6–14.

Levi, M., and Reuter, P. (2006) Money laundering. *Crime and Justice*, 34(1), 289–375.

Lewis, M. (2014) *Flashboys: Cracking the Money Code*. London: Allen Lane.

Leyshon, A. (1992) The transformation of regulatory order: Regulating the global economy and environment. *Geoforum*, 23(3), 249–267.

Leyshon, A., Lee, R., and Williams, C. C. (Eds.). (2003) *Alternative Economic Spaces*. London: SAGE.

Lilyblad, C. M. (2014) Illicit authority and its competitors: The constitution of governance in territories of limited statehood. *Territory, Politics, Governance*, 2(1), 72–93.

Lintner, B. (2005) Chinese organised crime. In M. Galeotti (Ed.), *Global Crime Today: The Changing Face of Organised Crime*. London: Routledge.

Lloyd, J. (2009) Investigative journalism, Saviano style. *Global Crime*, 10(3), 272–275.

Lombardo, R. M. (1997) The social organization of organized crime in Chicago. In P. J. Ryan and G. E. Rush (Eds.), *Understanding Organized Crime in Global Perspective: A Reader*. London: SAGE.

LSE Expert Group on the Economics of Drug Policy. (Ed.). (2014) *Ending the Drug Wars: Report of the LSE Expert Group on the Economics of Drug Policy*. London: London School of Economics.

Lunn, J. (Ed.). (2014) *Fieldwork in the Global South: Ethical Challenges and Dilemmas*. Abingdon, UK: Routledge.

Lupsha, P. (1983) Networks vs networking: analysis of an organised crime group. In G. Waldo (Ed.), *Career Criminals*. Beverly Hills, CA: SAGE.

Lupu, N. (2004) Towards a new articulation of alternative development: Lessons from coca supply reduction in Bolivia. *Development Policy Review*, 22(4), 405–421.

Lusthaus, J. (2013) How organised is organised cybercrime? *Global Crime*, 14(1), 52–60.

Mackinnon, D., and Cumbers, A. (2007) *An Introduction to Economic Geography: Globalization, Uneven Development and Place*. Harlow, UK: Pearson.

Madrazo Lajous, A. (2014) The constitutional costs of the "War on Drugs." In LSE Expert Group on the Economics of Drug Policy (Ed.), *Ending the Drug Wars: Report of the LSE Expert Group on the Economics of Drug Policy*. London: London School of Economics.

Madsen, F. (2009) *Transnational Organized Crime*. New York: Routledge.

Malm, A. E., Kinney, J. B., and Pollard, N. R. (2008) Social network and distance correlates of criminal associates involved in drug production. *Security Journal*, 21(1), 77–94.

Maltz, M. D. (1976) On defining organised crime: The development of a definition and a typology. *Crime and Delinquency*, 22(3), 338–346.

Manzo, K. (2005) Exploiting West Africa's children: Trafficking, slavery and uneven development. *Area*, 37(4), 393–401.

Marks, H. (1997) *Mr Nice: An Autobiography*. London: Vintage.

Marston, S. A., and Smith, N. (2001) States, scales and households: Limits to state thinking?: A response to Brenner. *Progress in Human Geography*, 25(4), 615–620.

Martin, C. (2015) Smuggling mobilities: Parasitic relations, and the aporetic openness of the shipping container. In T. Birtchnell, S. Savitsky, and J. Urry (Eds.), *Cargomobilities: Moving Materials in a Global Age*. New York: Routledge.

Martin, C. (2016) *Shipping Container*. London: Bloomsbury.

Massaro, V. (2015) The intimate entrenchment of Philadelphia's drug war. *Territory, Politics, Governance*, 3(4), 369–386.

McCann, B. (2006) The political evolution of Rio de Janeiro's favelas: Recent works. *Latin American Research Review*, 41(3), 149–163.

McCoy, A. W. (2004) The stimulus of prohibition: A critical history of the global narcotics trade. In M. K. Steinberg, J. J. Hobbs, and K. Mathewson (Eds.), *Dangerous Harvest: Drug Plants and the Transformation of Indigenous Landscapes*. Oxford, UK: Oxford University Press.

McMullan, J. (1984) *The Canting Crew: London's Criminal Underworld, 1550–1700*. New Brunswick, NJ: Rutgers University Press.

McNeill, D. (2004) *New Europe: Imagined Spaces*. London: Arnold.

Meehan, K. (2012) Water rights and wrongs: Illegality and informal use in Mexico and the US. In F. Sultana and A. Loftus (Eds.), *The Right to Water: Politics, Governance and Social Struggles*. Abingdon, UK: Earthscan.

Meehan, K., Shaw, I. G. R., and Marston, S. A. (2013, March) Political geographies of the object. *Political Geography*, 33, 1–10.

Mejia, D., and Restrepo, P. (2014) Why is strict prohibition collapsing?: A perspective from producer and transit countries. In LSE Expert Group on the Economics of Drug Policy (Ed.), *Ending the Drug Wars: Report of the LSE Expert Group on the Economics of Drug Policy*. London: London School of Economics.

Menkhaus, K. (2007) Governance without government in Somalia: Spoilers, state building, and the politics of coping. *International Security*, 31(3), 71–106.

Midgley, T., Briscoe, I., and Bertoli, D. (2014) *Identifying Approaches and Measuring Impacts of Programmes Focused on Transnational Organised Crime*. London: Department for International Development.

Monbiot, G. (2016) *How Did We Get into This Mess?* London: Verso.

Moore, A. W., and Perdue, N. A. (2014) Imagining a critical geopolitical cartography. *Geography Compass*, 8(12), 892–901.

Morselli, C. (2001) Structuring Mr. Nice: Entrepreneurial opportunities and brokerage positioning in the cannabis trade. *Crime, Law and Social Change*, 35(3), 203–244.

Morselli, C. (2005) *Contacts, Opportunities and Criminal Enterprise*. Toronto: University of Toronto Press.

Morselli, C. (2009) *Inside Criminal Networks*. New York: Springer.

Morton, J. (1992) *Gangland: London's Underworld*. London: Little, Brown.

Morton, J. (2001) *East End Gangland*. London: Warner.

Morton, J. (2008) *Gangland Soho*. London: Piatkus.

Moynagh, M., and Worsley, R. (2008) *Going Global: Key Questions for the 21st Century*. London: A. & C. Black.

Muggah, R. (2014) Deconstructing the fragile city: Exploring insecurity, violence and resilience. *Environment and Urbanization, 26*(2), 345–358.

Murdoch, J. (1997) Towards a geography of heterogeneous associations. *Progress in Human Geography, 21*(3), 321–337.

Murdoch, J. (1998) The spaces of actor–network theory. *Geoforum, 29*(4), 357–374.

Natarajan, M. (2000) Understanding the structure of a drug trafficking organization: A conversational analysis. In M. Natarajan and M. Hough (Eds.), *Illegal Drug Markets: From Research to Policy, Crime Prevention Studies: Volume 11*. Monsey, NY: Criminal Justice Press.

Natarajan, M. (2006) Understanding the structure of a large heroin distribution network: A quantitative analysis of qualitative data. *Journal of Quantitative Criminology, 22*(2), 171–192.

Naylor, R. (2002) *Wages of Crime: Black Markets, Illegal Finance, and the Underworld Economy*. Ithaca, NY: Cornell University Press.

Naylor, R. (2004) *Wages of Crime: Black Markets, Illegal Finance, and the Underworld Economy* (rev. ed.). Ithaca, NY: Cornell University Press.

Neal, S. (2010) Cybercrime, transgression and virtual environments. In J. Muncie, D. Talbot, and R. Walters (Eds.), *Crime: Local and Global*. Cullompton, UK: Devon.

Nelen, H. (2008) Real estate and serious forms of crime. *International Journal of Social Economics, 35*(10), 751–762.

Nordstrom, C. (2007) *Global Outlaws: Crime, Money, and Power in the Contemporary World*. Berkeley and Los Angeles: University of California Press.

Nordstrom, C. (2010) Women, economy, war. *International Review of the Red Cross, 92*(877), 161–176.

Nordstrom, C. (2011) Extra legality in the middle. *Middle East Review, 41*, 10–13.

Paglen, T., and Thompson, A. C. (2006) *Torture Taxi: On the Trail of the CIA's Rendition Flights*. Brooklyn, NY: Melville House.

Paley, D. (2015) *Drug War Capitalism*. Oakland, CA: AK Press.

Paoli, L. (2005) Italian organised crime: Mafia associations and criminal enterprises. In M. Galeotti (Ed.), *Global Crime Today: The Changing Face of Organised Crime*. London: Routledge.

Paoli, L., and Reuter, P. (2008) Drug trafficking and ethnic minorities in Western Europe. *European Journal of Criminology, 5*(1), 13–37.

Parry, B. (2008, September 15) *Amazon* (BBC television broadcast).

Passas, N. (2001) Globalization and transnational crime: Effects of criminogenic asymmetries. In P. Williams and D. Vlassis (Eds.), *Combating Transnational Crime: Concepts, Activities and Responses*. London: Frank Cass.

Peck, J., and Theodore, N. (2007) Variegated capitalism. *Progress in Human Geography, 31*(6), 731–772.

Pereira, L. (2010) Becoming coca: A materiality approach to commodity chain analysis of hoja de coca in Colombia. *Singapore Journal of Tropical Geography, 31*(3), 384–400.

Perrons, D. (2004) *Globalization and Social Change: People and Places in a Divided World*. London: Routledge.

Pham, J. P. (2011) Putting Somali piracy in context. In B. Hesse (Ed.), *Somalia: State Collapse, Terrorism and Piracy*. Abingdon, UK: Routledge.

Phillips, N. (2011) Informality, global production networks and the dynamics of "adverse incorporation." *Global Networks: A Journal of Transnational Affairs, 11*(3), 380–397.

Phillips, T. (2005) *Knockoff: The Deadly Trade in Counterfeit Goods*. London: Kogan Page.

Pickles, J. (2004) *A History of Spaces: Cartographic Reason, Mapping and the Geo-Coded World*. London: Routledge.

Pizzini-Gambetta, V. (2009) Women in *Gomorra*. *Global Crime, 10*(3), 267–271.

Potter, M. (2001) *Outlaws Inc.: Under the Radar and On the Black Market with the World's Most Dangerous Smugglers*. New York: Bloomsbury.

Pryce, K. (1979) *Endless Pressure: A Study of West Indian Lifestyles in Bristol*. London: Penguin.

Rafter, N. H. (2006) *Shots in the Mirror: Crime Films and Society*. Oxford, UK: Oxford University Press.

RAND Corporation. (2014) *How Big Is the U.S. Market for Illegal Drugs?* Santa Monica, CA: Author.

Ratcliffe, J. H. (2004) The hotspot matrix: A framework for the spatio-temporal targeting of crime reduction. *Police Practice and Research: An International Journal, 5*(1), 5–23.

Rengert, G. (1996) *The Geography of Illegal Drugs*. Boulder, CO: Westview Press.

Renzetti, C. M., and Lee, R. M. (Eds.). (1993) *Researching Sensitive Topics*. London: SAGE.

Reuter, P. (1985) *The Organization of Illegal Markets: An Economic Analysis*. New York: National Institute of Justice.

Reuter, P. (2014) The mobility of drug trafficking. In LSE Expert Group on the Economics of Drug Policy (Ed.), *Ending the Drug Wars: Report of the LSE Expert Group on the Economics of Drug Policy*. London: London School of Economics.

Reuter, P., and Haaga, J. (1989) *The Organization of High-Level Drug Markets: An Exploratory Study*. Santa Monica, CA: RAND Corporation.

Reuter, P., and Rubinstein, J. B. (1978) Fact, fantasy and organized crime. *Public Interest, 53*(1), 45–67.

Richards, I. (2016) "Flexible" capital accumulation in Islamic State social media. *Critical Studies on Terrorism, 9*(2), 205–225.

Risse, T. (2011) *Governance without a State?: Policies and Politics in Areas of Limited Statehood*. New York: Columbia University Press.

Roberts, S. (1995) Small place, big money: The Cayman Islands and the international financial system. *Economic Geography, 71*(3), 237–256.

Roberts, S. (1999) Confidence men: Offshore finance and citizenship. In M. P. Hampton and J. P. Abbott (Eds.), *Offshore Finance Centres and Tax Havens: The Rise of Global Capital*. Basingstoke, UK: Macmillan.

Robinson, J. (2002) *The Merger: The Conglomeration of International Organized Crime*. Woodstock, NY: Overlook Press.

Roy, A. (2011) Slumdog cities: Rethinking subaltern urbanism. *International Journal of Urban and Regional Research*, 35(2), 223–238.

Rozo, S. V., Gonzalez, V., Morales, C., and Soares, Y. (2015) Creating opportunities for rural producers: Impact evaluation of a pilot program in Colombia. *Journal of Drug Policy Analysis*. Retrieved from www.researchgate.net/publication/279711112_Creating_Opportunities_for_Rural_Producers_Impact_Evaluation_of_a_Pilot_Program_in_ColombiaDOI: 10.1515/jdpa-2014-0003.

Ruggiero, V. (2009) Transnational crime and global illicit economies. In E. Wilson (Ed.), *Government of the Shadows: Parapolitics and Criminal Sovereignty*. London: Pluto Press.

Ruggiero, V., and Khan, K. (2006) British South Asian communities and drug supply networks in the UK: A qualitative study. *International Journal on Drug Policy*, 17(6), 473–483.

Samers, M. (2005) The "underground economy" immigration and economic development in European Union: An agnostic-sceptic perspective. *International Journal of Economic Development*, 6(2), 199–272.

Santino, U. (2015) *Mafia and Antimafia*. London: IBS Tauris.

Saviano, R. (2008) *Gomorrah: Italy's Other Mafia*. London: Pan Books.

Saviano, R. (2009) Reply. *Global Crime*, 10(3), 276–279.

Saviano, R. (2016) *ZeroZeroZero*. London: Penguin.

Scalia, V. (2010) From the octopus to the spider's web: The transformations of the Scilian mafia under postfordism. *Trends in Organized Crime*, 13(4), 283–298.

Schoenmakers, Y. M. M., Bremmers, B., and Kleemans, E. R. (2013) Strategic versus emergent crime groups: The case of Vietnamese cannabis cultivation in the Netherlands. *Global Crime*, 14(4), 321–341.

Schwab, K., and Sala-i-Martín, X. (2015) *Global Competitiveness Report, 2015–2016*. Geneva, Switzerland: World Economic Forum.

Scott-Clark, C., and Levy, A. (2008, November 10) It's down your street and in your lane. *The Guardian Weekend* [UK].

Serious Organised Crime Agency. (2006) The United Kingdom threat assessment of serious organised crime. *Trends in Organized Crime*, 10(1), 52–57.

Shadoian, J. (2003) *Dreams and Dead Ends: The American Gangster Film*. Oxford, UK: Oxford University Press.

Sharman, J. C. (2011) *The Money Laundry: Regulating Criminal Finance in the Global Economy*. Ithaca, NY: Cornell University Press.

Sharpe, J. (1999) *Crime in Early Modern England, 1550–1750*. London: Longman.

Shelley, L. (2006) The globalization of crime and terrorism. *eJournal USA*, 11(1), 42–45.

Sheppard, E. (2002) The spaces and times of globalization: Place, scale, networks and positionality. *Economic Geography*, 78(3), 307–330.

Sheppard, E. (2008) Geographic dialectics? *Environment and Planning A*, 40(11), 2603–2612.

Sherry, F. (1986) *Raiders and Rebels*. New York: Hearst Marine Books.
Sidaway, J. (2007) Enclave space: A new metageography of development? *Area, 39*(3), 331–339.
Siegal, D. (2003) The transnational Russian mafia. In D. Siegel, H. van de Bunt, and D. Zaitch (Eds.), *Global Organized Crime: Trends and Developments*. Dordrecht, The Netherlands: Kluwer Academic.
Siegel, D., van de Bunt, H., and Zaitch, D. (Eds.). (2003) *Global Organized Crime: Trends and Developments*. Dordrecht, The Netherlands: Kluwer Academic.
Silverstone, D., and Savage, S. (2010) Farmers, factories and funds: Organised crime and illicit drugs cultivation within the British Vietnamese community. *Global Crime, 11*(1), 16–33.
Slade, G. (2007) The threat of the thief: Who has normative influence in Georgian society? *Global Crime, 8*(2), 172–179.
Smith, D. C. (1980) Paragons, pariahs and pirates: A spectrum based theory of enterprise. *Crime and Delinquency, 26*(3), 358–386.
Sproat, P. A. (2012) Phoney war or appeasement?: The policing of organised crime in the UK. *Trends in Organised Crime, 15*(4), 313–330.
Stedman-Jones, G. (1971) *Outcast London*. Oxford, UK: Oxford University Press.
Steger, M. B. (2003) *Globalization: A Very Short Introduction*. Oxford, UK: Oxford University Press.
Steinberg, M. K., Hobbs, J. J., and Mathewson, K. (2004) *Dangerous Harvest: Drug Plants and the Transformation of Indigenous Landscapes*. New York: Oxford University Press.
Stelfox, P. (1996) *Gang Violence: Strategic and Tactical Options*. London: Home Office Police Research Group.
Stephenson, S. (2015) *Gangs of Russia: From the Streets to the Corridors of Power*. Ithaca, NY: Cornell University Press.
Sterling, C. (1994) *Crime without Frontiers: The Worldwide Expansion of Organised Crime and the Pax Mafiosa*. London: Warner Books.
Stewart, H. (2012, July 22) Revealed: Wealth doesn't trickle down—it floods offshore. *The Observer* [UK].
Swain, A., Mykhnenko, V., and French, S. (2010) The corruption industry and transition: Neoliberalizing post-Soviet space. In K. Birch and V. Mykhnenko (Eds.), *The Rise and Fall of Neoliberalism: The Collapse of an Economic Order?* London: Zed Books.
Syal, R. (2009, December 13) Drug money "saved the banks" in global crisis: UN adviser says institutions effectively laundered $352bn in criminal proceeds. *The Observer* [UK].
Symantec. (2011) *Norton Cybercrime Report 2011*. Retrieved May 5, 2017, from http://now-static.norton.com/now/en/pu/images/Promotions/2012/cybercrime/assets/downloads/en-us/NCR-DataSheet.pdf.
Taylor, J. S., Jasparro, C., and Mattson, K. (2013) Geographers and drugs: A survey of the literature. *Geographical Review, 103*(3), 415–430.
Taylor, R. B. (2003) Crime prevention through environmental design (CPTED): Yes, no, maybe, unknowable, and all of the above. In R. B. Bechtel and A.

Churchman (Eds.), *Handbook of Environmental Psychology*. New York: Wiley.

Tilly, C. (1985) War making and state making as organized crime. In P. Evans, D. Rueschemeyer, and T. Skocpol (Eds.), *Bringing the State Back In*. Cambridge, UK: Cambridge University Press.

Tobacco Manufacturers Association. (2016) Cigarette Prices across Europe. Retrieved May 9, 2017, from *http://the-tma.org.uk/wp-content/uploads/2016/12/AIT_EU_MAP_CG_2016.jpg*.

Transcrime. (2015) *From Illegal Markets to Legitimate Businesses: The Portfolio of Organised Crime in Europe*. Milan: Joint Research Centre on Transnational Crime.

Truman E. M., and Reuter P. (2004) *Chasing Dirty Money: Progress on Anti-Money Laundering*. Washington, DC: Institute for International Economics.

Tuathail, G. (1996) *Critical Geopolitics: The Politics of Writing Global Space*. London: Routledge.

Uchtenhagen, A. (2009) Heroin-assisted treatment in Switzerland: A case study in policy change. *Addiction, 105*(1), 29–37.

Unger, B., and Rawlings, G. (2008) Competing for criminal money. *Global Business and Economics Review, 10*(3), 331–352.

United Nations Centre for International Crime Prevention. (2002) *Towards a Monitoring System for Transnational Organized Crime Trends: Results of a Pilot Survey of 40 Selected Organized Crime Groups in 16 Countries*. Vienna: United Nations.

United Nations Office on Drugs and Crime. (2007) *Cocaine Trafficking in West Africa: The Threat to Stability and Development*. Vienna: Author.

United Nations Office on Drugs and Crime. (2008) *World Drugs Report 2008*. Vienna: Author.

United Nations Office on Drugs and Crime. (2010a) *The Globalization of Crime: A Transnational Organized Crime Threat Assessment*. Vienna: Author.

United Nations Office on Drugs and Crime. (2010b) *World Drugs Report 2010*. Vienna: Author.

United Nations Office on Drugs and Crime. (2011) *Estimating Illicit Financial Flows Resulting from Drug Trafficking and Other Transnational Organized Crime: Research Report*. Vienna: Author.

United Nations Office on Drugs and Crime. (2012) *Opiate Flow through Northern Afghanistan and Central Asia: A Threat Assessment*. Vienna: Author.

United Nations Office on Drugs and Crime. (2014a) *Global Report on Trafficking in Persons 2014*. Vienna: Author.

United Nations Office on Drugs and Crime. (2014b) *World Drugs Report 2014*. Vienna: Author.

United Nations Office on Drugs and Crime. (2015a) *Drug Money: The Illicit Proceeds of Opiates Trafficked on the Balkan Route*. Vienna: Author.

United Nations Office on Drugs and Crime. (2015b) Transnational organized crime—Let's put them out of business. Retrieved May 10, 2017, from *www.unodc.org/toc*.

United Nations Office on Drugs and Crime. (2015c) *World Drugs Report 2015*. Vienna: Author.
United Nations Office on Drugs and Crime. (2016) *World Drugs Report 2016*. Vienna: Author.
United Nations Office on Drugs and Crime. (n.d.) Transnational organized crime—The globalized illegal economy. Retrieved May 10, 2017, from *www.unodc.org/documents/toc/factsheets/TOC12_fs_general_EN_HIRES.pdf*.
Urry, J. (2014) *Offshoring*. Cambridge, UK: Polity Press.
van de Bunt, H., and Siegel, D. (2003) Introduction. In D. Siegel, H. van de Bunt, and D. Zaitch (Eds.), *Global Organized Crime: Trends and Developments*. Norwell, MA: Kluwer Academic.
van Duyne, P. C. (1996) *Organized Crime in Europe*. New York: Nova Science.
van Schendel, W. (2005) Spaces of engagement: How borderlands, illegal flows and territorial states interlock. In W. van Schendel and I. Abraham (Eds.), *Illicit Flows and Criminal Things: States, Borders and the Other Side of Globalization*. Bloomington: Indiana University Press.
van Schendel, W., and Abraham, I. (Eds.). (2005) *Illicit Flows and Criminal Things: States, Borders and the Other Side of Globalization*. Bloomington: Indiana University Press.
Vander Beken, T., Savona, E., Korsell, L., Defruytier, M., Di Nicola, A., Heber, A., et al. (2005) *Measuring Organised Crime in Europe: A Feasibility Study of a Risk Based Methodology Across the European Union*. Antwerp-Apeldoorn: Maklu.
van Dijk, J. (2007) Mafia markers: Assessing organized crime and its impact upon societies. *Trends in Organized Crime*, 10(1), 39–56.
Varese, F. (2009) The Camorra closely observed. *Global Crime*, 10(3), 262–266.
Varese, F. (2011) *Mafias on the Move: How Organized Crime Conquers New Territories*. Princeton, NJ: Princeton University Press.
von Lampe, K. (2006) The interdisciplinary dimensions of the study of organized crime. *Trends in Organized Crime*, 9(3), 77–95.
Vulliamy, E. (2010) *Amexica: War Along the Borderland*. London: Bodley Head.
Vulliamy, E. (2012, July 7) We revealed the links between banks and drugs last year. Nothing has changed. *The Observer* [UK].
Wainwright, T. (2016) *Narco Nomics: How to Run a Drug Cartel*. London: Ebury Press.
Walker, J., and Unger, B. (2009) Measuring global money laundering: The Walker gravity model. *Review of Law and Economics*, 5(2), 821–853.
Wall, D. S., and Large, J. (2010) Jailhouse frocks: Locating the public interest in policing counterfeit luxury fashion goods. *British Journal of Criminology*, 50(6), 1094–1116.
Walters, R. (2010) Eco crime. In J. Muncie, D. Talbot, and R. Walters (Eds.), *Crime: Local and Global*. Cullompton, UK: Devon.
Watt, P., and Zepeda, R. (2012) *Drug War Mexico: Politics, Neoliberalism and Violence in the New Narcoeconomy*. London: Zed Books.
Weinstein, L. (2008) Mumbai's development mafias: Globalization, organized

crime and land development. *International Journal of Urban and Regional Research, 32*(1), 22–39.
Westmarland, L. (2010) Gender abuse and people trafficking. In J. Muncie, D. Talbot, and R. Walters (Eds.), *Crime: Local and Global*. Cullompton, UK: Devon.
Williams, C. C. (2004) *Cash-in-Hand Work: The Underground Sector and the Hidden Economy of Favours*. Basingstoke, UK: Palgrave Macmillan.
Williams, C. C. (2006) *The Hidden Enterprise Culture: Entrepreneurship in the Underground Economy*. Cheltenham, UK: Edward Elgar.
Williams, P. (2012) *Badfellas*, London: Penguin.
Williams, P. J. (2001) Crime, illicit markets, and money laundering. In P. J Simmons and C. Ouderen (Eds.), *Challenges in International Governance*. Washington, DC: Carnegie Endowment.
Williams, T. (1989) *The Cocaine Kids*. Reading, MA: Addison-Wesley.
Williams, T. (1992) *Crack House*. Reading, MA: Addison-Wesley.
Wilson, E. (2009) Deconstructing the shadows. In E. Wilson (Ed.), *Government of the Shadows: Parapolitics and Criminal Sovereignty*. London: Pluto Press.
Wiltshire, S., Bancroft, A., Amos, A., and Parry, O. (2001) "They're doing the people a service": Qualitative study of smoking, smuggling and social deprivation. *British Medical Journal, 323*, 203–207.
Wiseman, T., and Walker, P. (2017) U.S. interstate underground trade flow: A gravity model approach. *Review of Law and Economics, 13*(2). [Epub ahead of print]
Wójcik, D. (2013) The dark side of NY–LON: Financial centres and the global financial crisis. *Urban Studies, 50*(13), 2736–2752.
Wood, D., with Fels, J., and Krygier, J. (2010) *Rethinking the Power of Maps*. New York: Guilford Press.
Woodiwiss, M. (1988) *Crime, Crusades and Corruption: Prohibitions in the United States, 1900–1987*. London: Pinter.
Woodiwiss, M. (2001) *Organized Crime and American Power*. Toronto: University of Toronto Press.
Woodiwiss, M. (2003) Transnational organised crime: The global reach of an American concept. In A. Edwards and P. Gill (Eds.), *Transnational Organised Crime: Perspectives on Global Security*. London: Routledge.
Woodiwiss, M. (2012) The past and present of transnational organized crime in America. In F. Allum and S. Gilmor (Eds.), *The Routledge Handbook of Transnational Organized Crime*. Abingdon, UK: Routledge.
Woodiwiss, M., and Hobbs, D. (2009) Organized evil and the Atlantic challenge: Moral panics and the rhetoric of organized crime policing in America and Britain. *British Journal of Criminology, 49*(1), 106–125.
World Economic Forum. (2011) *Global Risks 2011*. New York: Author.
World Economic Forum. (2012) *Organized Crime Enablers*. Geneva, Switzerland: Author.
Wright, A. (2006). *Organised Crime*. Uffculme, UK: Willan.
Wright, C. (2004) Tackling conflict diamonds: The Kimberley Process Certification Scheme. *International Peacekeeping, 11*(4), 697–708.

Wright, M. W. (2011) Necropolitics, narcopolitics, and femicide: Gendered violence on the Mexico–US border. *Signs, 36*(3), 707–731.

Yanik, L. K. (2009) The metamorphosis of metaphors of vision: "Bridging" Turkey's location, role and identity after the end of the Cold War. *Geopolitics, 14*(3), 531–549.

Zhang, S., and Chin, K.-L. (2004) *Characteristics of Chinese Human Smugglers*. Washington, DC: U.S. Department of Justice.

Zook, M. A. (2003) Underground globalization: Mapping the space of flows of the Internet adult industry. *Environment and Planning A, 35*(7), 1261–1286.

INDEX

Note. *f* or *t* following a page number indicates a figure or a table.

Academic perspectives, 6–7, 66–69. *See also* Research
Actor–network theory, 64
Adaptiveness, 86
Afghanistan
　criminal commodity movements and, 130, 131, 138
　criminal organization and, 89–90
　drug economies and, 24
　networked criminal mobilities and, 126–127
　organized criminal markets and, 36
　responding to organized crime and, 163
Africa
　criminal commodity movements and, 134
　economies of organized crime and, 21
　human trafficking economies and, 30–31
　responding to organized crime and, 177
Alaska, 67–68
Angola, 35
Anonymity, 102–106
Anti-organized crime efforts, 154–156, 155*f*, 160–161, 173–179. *See also* Responding to organized crime
Archival analysis, 61*t*
Arms trafficking. *See* Firearms trafficking
Asymmetry
　global economy and, 102–106
　network ontologies and, 120
ATM skimming, 112
Australia, 170

B

Balloon effect
　drug economies and, 25
　responding to organized crime and, 159
Baltic states, 31–32
Bank robberies, 73

Banking, 106–110. *See also* Financial contexts
Belgium, 136
Birmingham's Jewellery Quarter, 94
Bolivia
　drug economies and, 25
　responding to organized crime and, 159
Border Wars (television series), 66
Borderland regions
　local geographical contexts and, 114–115
　responding to organized crime and, 163
Borders and border security
　drug economies and, 158
　local geographical contexts and, 113*t*, 114–115
　state weakness and, 163
Brazil
　cybercrime economies and, 31–32
　responding to organized crime and, 166
Bribery
　regulating criminal economies and, 105
　responding to organized crime and, 158–159
Britain, 64
British Columbia
　drug economies and, 36–37
　local geographical contexts and, 116
Brussels, 136
Business Environment and Economic Performance Survey (Homes, 2016), 50
Businesses, 13

C

Camorra, 128
Canada
　decriminalization and legalization movements and, 171
　local geographical contexts and, 116

213

Cannabis
  criminal organization and, 82, 89–90
  decriminalization and legalization movements and, 171–172
  economies of, 23–24, 25, 26f, 28, 36–37
  local geographical contexts and, 112
  networked criminal mobilities and, 127–128
  production of, 28
  regulating criminal economies and, 93, 106
  See also Drugs
Cargo ships, 120–121
Caribbean, 159
Cartographical analysis, 17
Categorization of phenomenon, 145–150, 149f
Causality, 165–166
Cayman Islands, 108
Cellular group organization, 73–74
Central America, 94
Central Asia, 138–139
Change in economic organization, 71
Change theories, 154, 155f
China
  counterfeiting economies and, 29–30
  criminal organization and, 81
  cybercrime economies and, 31–32
  local geographical contexts and, 116
  manufacturing and, 187–188
  networked criminal mobilities and, 128
  networked group organization and, 87
  regulating criminal economies and, 102–103
  spatialities of organized crime, 99
Cigarettes. *See* Tobacco products, illicit trade in
Climatic factors, 25
Cocaine
  criminal organization forms and, 75
  economies of, 24, 25, 27–28, 36–37
  investigative journalism and, 55–56
  local geographical contexts and, 112
  networked criminal mobilities and, 127–128
  production of, 27–28
  See also Drugs
Coke, Christopher "Dudus," 76
Colombia
  drug economies and, 24, 25, 36–37
  multiscalar approaches and, 117
  regulating criminal economies and, 105
  responding to organized crime and, 158, 159, 161, 163
Colorado, 67–68
Coltan, transit of, 127–128
Commercial crimes, 5
Commodity chain
  criminal commodity movements and, 130–137
  networked criminal mobilities and, 127
Commodity movements, 137–150, 144f, 145t, 149f

Composite Organized Crime Index (COCI), 35–36, 51, 52f
Consumption
  criminal mobilities and, 125
  drug economies and, 157–158
  responding to organized crime and, 167–168
Container Security Initiative (CSI), 104–105
Container ships, 103–105
Contemporary economic geography, 9–13, 72–73, 184. *See also* Economic geography
Contemporary organized crime, 21–23
Contingent regulatory landscapes, 167
Convention on International Trade in Endangered Species of Wild Fauna and Flora (CITES), 167
Convention on Psychotropic Substances (1971), 157
Conventions, 167
Corruption
  counterfeiting and, 87
  economic geography perspective and, 18
  economies of, 22
  poly-crime groups and, 76
Cosa Nostra, 89–90
Counterfeiting
  benefits from organized crime and, 101–102
  economies of, 22, 23t, 28–30
  networked group organization and, 87
  overview, 187–188
  poly-crime groups and, 76
  regulating criminal economies and, 105
  spatialities of organized crime, 99
  See also Intellectual property right theft
Criminal commodity movement, 137–150, 144f, 145t, 149f
Criminal economies
  counterfeiting economies and, 29–30
  criminal organization and, 97
  cybercrime economies and, 31–33
  drug economies and, 23–28, 23t, 26f
  economies of organized crime and, 21–23
  environmental crime economies and, 33–34
  global economy and, 39–40
  human trafficking economies and, 30–31
  local geographical contexts and, 113–116
  money laundering economies and, 34
  multiscalar approaches, 117–118
  networked criminal mobilities and, 127–128
  organized criminal markets and, 35–38
  overview, 21–22, 40
  regulating, 92–96
  responding to organized crime and, 153–154, 163, 174, 177–178
  spatialities of organized crime, 121–122
  See also Economies of organized crime

Criminal mobilities
   criminal commodity movements and, 130–137
   networked criminal mobilities, 126–130
   overview, 123–126, 150–151
   representations of illicit commodity movements, 137–150, 144f, 145t, 149f
   responding to organized crime and, 162–163
   See also Mobility
Criminal organization
   forms of, 73–77
   overview, 71–73, 96–97
   under post-Fordism, 77–88
   regional differences and, 88–92
   regulating criminal economies and, 92–96
Criminalization of drugs, 168–169. *See also* Drugs; War on Drugs
Criminology perspective, 127–128
Critical geopolitics, 65–66
Cultural factors
   criminal organization and, 84, 92
   definitions of organized crime and, 3–4
   local geographical contexts and, 113t
   narcotics and, 16–17
   organized criminal markets and, 36, 39–40
   overview, 6–7, 10
   responding to organized crime and, 163–164
Cybercrime
   economies of, 22, 31–33
   local geographical contexts and, 112
   networked criminal mobilities and, 128
   responding to organized crime and, 155–156

D

Dangerous Harvest (Steinberg, Hobbs, and Mathewson, 2004), 16–17
Data collection, 69. *See also* Research
Decapitation approaches, 86
Decriminalization, 169–175
Delaware, 108
Democratic Republic of the Congo, 35
Demographic factors
   criminal organization and, 82
   research and, 186–187
Developing nations, 161
Diamond exchange, 167, 168
Diaspora populations, 80
Distribution of costs, 161
Diverse economies, 12
Diversity in economic organization, 71
"Drug flow," 143, 144f, 145t
Drug-related activities, 17–18
Drugs
   criminal commodity movements and, 130–137, 138–139, 140–145, 144f, 145t, 147–150
   criminal organization and, 82, 89–90
   criminal organization forms and, 75
   decriminalization and legalization movements and, 169–175
   economies of, 22, 23–28, 23t, 26f, 36–37, 45–46
   historical and ethnographic accounts and, 63
   investigative journalism and, 55–56
   local geographical contexts and, 112, 115–116
   networked criminal mobilities and, 126–127, 128–130
   policymaking and, 67–68, 169–175
   poly-crime groups and, 76
   regulating criminal economies and, 93, 105–106
   responding to organized crime and, 160, 161–162, 165, 168–169
   social media and online sources, 58–59
   See also Narcotics; Prohibition policies; War on Drugs
Durban, 134–135

E

Eastern Europe, 31–32
Economic change, 19
Economic development, 165, 166
Economic geography
   global economy and, 39–40
   organized criminal markets and, 39–40
   overview, 1–3, 2f, 9–13, 17–18, 40, 71–73, 181–188
   responding to organized crime and, 160–161, 173
   spatialities of organized crime, 98–100
Economic growth, 164–165
Economic marginalization, 162–163
Economic markets associated with organized crime, 21–23
Economic rationality, 76
Economic transformations, 83–84
Economies of organized crime. *See also* Criminal economies
   criminal organization and, 78–79
   overview, 12–13, 21–23, 44–45
   research and, 44–45, 69
Ecstasy (MDMA) trafficking, 129–130. *See also* Drugs
Ecuador, 159
Employment
   drug economies and, 36–37
   responding to organized crime and, 165–166
Environmental crime
   economies of, 33–34
   poly-crime groups and, 76
   responding to organized crime and, 167
Environmental factors
   drug economies and, 25
   local geographical contexts and, 113t
Equilibrium, 159–160
Eradication, 144

Index

Ethnicity
  criminal organization and, 79–82, 89–90
  global economy and, 39
  regulating criminal economies and, 93
Ethnographic accounts, 60–65, 61t–62t, 62t, 70, 182
Europe
  counterfeiting economies and, 29
  criminal commodity movements and, 130
  cybercrime economies and, 31–32
  decriminalization and legalization movements and, 171
  drug economies and, 28
  human trafficking economies and, 30–31
  organized criminal markets and, 36
  policymaking and, 67–68
  responding to organized crime and, 154
  tax havens and, 108
European Union
  criminal organization forms and, 76
  economies of organized crime and, 22, 29–30, 30–31
  responding to organized crime and, 168
Exchange, 125
Experience surveys, 49–50
Extortion, 76

Family ties
  criminal organization and, 89–90
  regulating criminal economies and, 93
  See also Kinship
Financial Action Task Force, 160–161, 167
Financial contexts
  benefits from organized crime and, 102
  criminal commodity movements and, 139–140
  global economy and, 106–110
  networked criminal mobilities and, 127–128
  See also Banking
Firearms trafficking, 22, 23t, 28–30
Flexibility, 71, 86
Flows, metaphors of, 140–145, 144f, 145t
Follow-the-thing ethnographies, 64, 124. *See also* Ethnographic accounts
Foreign mobsters, 113t
Formal capitalist economies, 1–3, 2f
Franchising, 94–95
Fraud, 22
Frozen River (film), 66

G

Gambling, illegal, 76
Gangster autobiographies, 56–58
Gangsterism, 6, 163
Geographical contexts
  local geographical contexts, 110–116, 113t
  overview, 16–18
  responding to organized crime and, 175–179
Geopolitical factors, 25, 27

Georgia
  organized criminal markets and, 36
  responding to organized crime and, 163
Ghana, 177
Glenny, Misha, 55, 58
Global capitalism, 78
Global Competitiveness Report (Schwab and Sala-i-Martín, 2015), 50
Global contexts
  networked group organization and, 86–89
  responding to organized crime and, 154, 155f, 160–161, 167
  See also Global economy
Global economy
  counterfeiting economies and, 29–30
  criminal commodity movements and, 135–136, 148–149
  criminal organization in, 39–40
  cybercrime economies and, 31–33
  drug economies and, 23–28, 23t, 26f
  economies of organized crime and, 21–23
  environmental crime economies and, 33–34
  human trafficking economies and, 30–31
  local geographical contexts and, 110–116, 113t
  money laundering economies and, 34
  multiscalar approaches, 117–118
  network ontologies and, 120
  organized criminal markets and, 35–38
  overview, 21–22, 40, 100–110, 182–183
  regulating criminal economies and, 102–103
  responding to organized crime and, 178
Global Illicit Drug Trends, 147
Global mobilities. *See* Criminal mobilities; Mobility
Global production networks (GPN) approach
  criminal commodity movements and, 133–134, 136
  network ontologies and, 120
Global regulations, 167. *See also* Regulation
Global value chains, 136
Globalization
  criminal commodity movements and, 130–137
  criminal organization and, 80–82
  historical and ethnographic accounts and, 64
  overview, 184
  regulating criminal economies and, 103–104
  See also Spatial patterns
Gomorrah (Saviano, 2008), 55, 99

H

Harm reduction, 174
Hernàndez, Anabel, 54–55, 58
Heroin, 24–25, 27–28, 46. *See also* Drugs
Hierarchical organizational structures, 6
Historical accounts, 60–65, 61t–62t
Historical/archival analysis, 61t

Hong Kong, 105
Human trafficking
　economies of, 22, 23t, 30–31
　responding to organized crime and, 154

I

Idolization of criminality, 113t
Illegal activities
　geographical literature on organized crime and, 16–18
　overview, 14–15
　political and social views of, 13–15
Illegal capitalist economies, 1–3, 2f
Illegal drugs. See Drugs
Illegal migrants, 140
Illegal or illicit services
　geographical literature on organized crime and, 16–18
　overview, 4–5
　political and social views of, 13–15
　regulating criminal economies and, 102–106
Illicit commodities, 87
Income
　drug economies and, 36–37
　responding to organized crime and, 165–166
India, 26f
Indigenous peoples, 16–17
Informal economies, 1–3, 2f
Infrastructure provision, 165
Inherent paradox, 159–160
Innovative Marketing, 73–74
Intellectual property right theft
　economies of, 29
　networked group organization and, 87
　See also Counterfeiting
International policing, 87–88, 158–159, 167. See also Law enforcement; Policing perspectives
Interventions, 174
Interviewing method, 62t
Intimidation, 105
Investigative journalism sources, 53–56
Islamic State, 32
Israel, 89–90
Italy
　criminal organization and, 74–75, 81
　local geographical contexts and, 116
　networked criminal mobilities and, 128
　organized crime indexes, 51
　poly-crime groups and, 76
　responding to organized crime and, 163

J

Jamaica
　local geographical contexts and, 114–115
　poly-crime groups and, 76
　responding to organized crime and, 166
Jammu, 26f

Japan
　criminal commodity movements and, 133–134, 136
　investigative journalism and, 55
　local geographical contexts and, 116
　poly-crime groups and, 76

K

Kashmir, 26f
Kimberley process for the certification of diamonds, 167, 168
Kingston, 114–115
Kinship
　criminal organization and, 79–82
　regulating criminal economies and, 93
　See also Family ties
Kray twins, 57

L

Labor, illegal or trafficked
　counterfeiting and, 87
　networked criminal mobilities and, 126–127
　See also Human trafficking
Laboratories, 120–121
Lao People's Democratic Republic, 24
Large corporation criminal organization form, 74
Latin America
　counterfeiting economies and, 29
　drug economies and, 25
　human trafficking economies and, 30–31
　spatialities of organized crime, 98
Latvia, 31–32
Law enforcement
　criminal commodity movements and, 139–140, 145
　decriminalization and legalization movements and, 172
　international policing and, 87–88
　local geographical contexts and, 113–116, 113t
　networked group organization and, 86, 87
　overview, 47, 162–163
　state weakness and, 162–163
　threat and risk assessments and, 53
　See also Policing perspectives; Responding to organized crime
Legal activities
　geographical literature on organized crime and, 16–18
　overview, 14–15
　political and social views of, 13–15
Legalization, 169–175
Legislation, 14–15, 153. See also Policymaking; Prohibition policies
Liberalization, 166
Libya, 154

Licit activities
  criminal commodity movements and, 142–143
  geographical literature on organized crime and, 16–18
  overview, 14–15
  political and social views of, 13–15
Licit economy
  benefits from organized crime and, 101–102
  criminal organization and, 78–79, 81
  networked group organization and, 85
  overview, 184
  regulating criminal economies and, 102–104
Local geographic contexts
  network ontologies and, 119–120
  overview, 110–116, 113*t*
  responding to organized crime and, 154, 155*f*
  state weakness and, 162–163
Local protectionism, 14–15
London, 108
Los Señores del Narco (Hernàndez, 2010), 54
Luxembourg, 108

M

Madoff, Bernard, 57
Mafia associations
  criminal organization forms and, 74–75
  local geographical contexts and, 112, 116
  poly-crime groups and, 76
  regulating criminal economies and, 93–94
Mafia Index, 51
Mapping of organized crime
  criminal commodity movements and, 145–150, 149*f*
  organized crime indexes, 50–53, 52*f*
  overview, 184–185
Marginalized populations, 161, 164–165
Marginalized urban spaces, 64
Marijuana. *See* Cannabis
Maritime piracy
  local geographical contexts and, 112
  responding to organized crime and, 166
Market economies
  benefits from organized crime and, 102
  networked criminal mobilities and, 129–130
Market equilibrium, 159–160
Markets associated with organized crime. *See* Economic markets associated with organized crime
Material accumulation, 76–77
McMafia (Glenny, 2008), 55, 56
Measuring organized crime
  investigative journalism sources, 53–56
  organized crime indexes, 50–53, 52*f*
  overview, 41–49, 68–70
  perception and experience surveys, 49–50
  policymaking and, 66–68
  threat and risk assessments, 53
  See also Research
Media, 47, 58–59
Methodology
  historical and ethnographic accounts and, 61*t*–62*t*
  investigative journalism and, 55–56
  organized crime indexes, 49–50
  overview, 41–42, 183
  See also Measuring organized crime; Research
Mexico
  criminal organization forms and, 74–75
  drug economies and, 24–25, 27
  regulating criminal economies and, 92–93, 94
  responding to organized crime and, 158–159, 160, 163, 166
  social media and online sources, 58–59, 60
México en Llamas (Hernàndez, 2013), 54–55
Migration, 80–82. *See also* Mobility; Spatial patterns
Military intervention, 158
Mobility
  criminal organization and, 80–82
  global economy and, 102–106
  networked group organization and, 85
  overview, 150–151
  regulating criminal economies and, 93–95, 106
  responding to organized crime and, 153–154, 159
  See also Criminal mobilities; Globalization; Migration; Spatial patterns
Money laundering
  benefits from organized crime and, 101–102
  counterfeiting and, 87
  economies of, 34
  overview, 44–45, 181–182
  poly-crime groups and, 76
  responding to organized crime and, 160–161, 167
Monitoring, 172–173
Moral commitment, 76–77
Moral geography, 16–17
Mozambique, 133–135, 136
Multicultural factors, 80–82. *See also* Cultural factors
Multinational corporation criminal organization form, 74
Multiple criminal markets
  criminal organization forms and, 76
  responding to organized crime and, 155–156
Multiscalar approaches, 117–118, 121–122, 185, 186
Myanmar, 24

N

Naples, 128
Narcotics
  benefits from organized crime and, 101–102
  criminal commodity movements and, 130–137
  criminal organization and, 75, 89–90
  decriminalization and legalization movements and, 169–175
  environmental crimes and, 33–34
  geographical literature on organized crime and, 16–17
  local geographical contexts and, 115–116
  policymaking and, 67–68
  regulating criminal economies and, 105–106
  responding to organized crime and, 167
  spatialities of organized crime, 99
  See also Drugs
National contexts, 154, 155*f*, 167
National security, 48
'Ndranghetta, 89–90
Netherlands
  criminal organization and, 80, 83
  decriminalization and legalization movements and, 170
  networked criminal mobilities and, 127–128
  regulating criminal economies and, 93
Network analysis, 61*t*
Network ontologies, 119–121, 178–179
Networked group organization
  network ontologies and, 119–121
  networked criminal mobilities, 126–130
  overview, 73–74, 85–89, 97, 184–185
New Jersey, 98–99
New York, 108
New Zealand, 170
Nigeria, 31–33, 35
North Korea
  cybercrime economies and, 32
  regulating criminal economies and, 105

O

Offshore tax havens. *See* Tax havens
Online sources, 58–60
Opacity in economic systems
  global economy and, 102–106
  network ontologies and, 120–121
  overview, 12
Operation Condor, 75
Operation Sophia, 154
Opiates
  criminal commodity movements and, 130–131
  criminal organization and, 92
  economies of, 24–25, 26*f*, 45–46
  local geographical contexts and, 112
  See also Drugs

Organisation for Economic Co-operation and Development (OECD)
  economies of organized crime and, 22
  tax havens and, 108–110
Organizational determinism, 82–83
Organizational structures
  local geographical contexts and, 112
  overview, 71–73
  regional differences in, 88–92
  state weakness and, 113–116
Organized crime in general
  global economy and, 39–40
  overview, 3–9, 181–188
  the state and, 35–36
Organized crime indexes, 50–53, 52*f*

P

Pakistan
  criminal commodity movements and, 134
  drug economies and, 26*f*
Patterns of organized crime, 22–23, 35–38. *See also* Spatial patterns
Pax mafiosa, 81
People trafficking. *See* Human trafficking
Perception surveys, 49–50
Peru
  drug economies and, 25
  responding to organized crime and, 159
Pharmaceutical drug counterfeiting, 29. *See also* Counterfeiting; Drugs
Piedmont, 116
Poached game, 33. *See also* Environmental crime
Policing perspectives
  criminal mobilities and, 151
  networked group organization and, 87–88
  overview, 6–7, 46–48
  See also Law enforcement
Policy pluralism, 174
Policymaking
  criminal commodity movements and, 130–131
  geographical contexts and, 175–179
  global economy and, 177–178
  local geographical contexts and, 116
  overview, 169–175, 179–180, 188
  research and, 66–68, 69
  responding to organized crime and, 160–161
  See also Prohibition policies; Responding to organized crime; War on Drugs
Political contexts
  criminal organization and, 91–92
  overview, 6–7, 10, 47
  production of organized crime and, 13–15
  regulating criminal economies and, 102–103
  responding to organized crime and, 163–164
  spatialities of organized crime, 98–99
Political embeddedness, 74–75

Political prisoners, 139–140
Political-economy perspective, 10–11
Poly-crime groups, 76
Portugal
  decriminalization and legalization movements and, 170
  policymaking and, 67
Positionality, 120
Post-Fordism, 77–88
Post-Soviet space, 35–36
Predatory crimes, 4–5
Prison systems, 91
Product scarcity, 160
Production
  criminal mobilities and, 125
  destruction of production sites, 158
  responding to organized crime and, 159
Prohibition policies
  criminal commodity movements and, 130–131
  decriminalization and legalization movements and, 169–175
  local geographical contexts and, 113t
  overview, 153, 157–158, 174, 179–180, 185
  responding to organized crime and, 167
  See also Drugs; Policymaking
Property development, 76
Prostitution, 76
Protection
  local geographical contexts and, 113t
  poly-crime groups and, 76
Puerto Rico, 108
Punitive responses, 169–170. See also Responding to organized crime

R
Reductionism, 76
Reformist discourse, 169–175
Regional contexts
  criminal mobilities and, 125
  criminal organization and, 80–82
  network ontologies and, 119–120
  organizational structures and, 88–92
  organized criminal markets and, 35–38
  responding to organized crime and, 154, 155f, 164, 165–166, 167, 178–179
Regulation
  criminal economies and, 92–96
  decriminalization and legalization movements and, 172
  global economy and, 102–106
  responding to organized crime and, 167
Remoteness, geographical, 106
Reports, official, 145–150, 149f
Reputations, 94
Research
  autobiographies and, 56–58
  critical geopolitics, 65–66
  future directions for, 186–187

  historical and ethnographic accounts, 60–65, 61t–62t
  investigative journalism sources, 53–56
  organized crime indexes, 50–53
  overview, 41–43, 68–70, 183
  perception and experience surveys, 49–50
  policymaking and, 66–68
  responding to organized crime and, 164–166, 173, 174
  social media and online sources, 58–60
  See also Academic perspectives; Measuring organized crime
Resource curse, 35
Responding to organized crime
  decriminalization and legalization movements and, 169–175
  extant approaches, 156–169
  overview, 152–156, 155f, 179–180, 183, 185
  See also Law enforcement; Policing perspectives; Policymaking
Retail theft, 127–128
Risk assessments, 53
Risk in economic systems, 12
Ritualistic representation, 84
Rotterdam, 104
Royal United Services Institute, 67
Rumors, 94
Russia
  counterfeiting economies and, 29
  criminal organization and, 77, 81, 83–84, 89–90, 91–92
  cybercrime economies and, 31–32
  multiscalar approaches and, 117–118
  organized criminal markets and, 35–36
  spatialities of organized crime, 99

S
Saviano, Roberto, 55–56, 58, 128
Scalar multiplicity, 155–156
Secondhand goods, 139–140
Sex trafficking. See Human trafficking
Sex workers, 170
Sexual exploitation, 30–31. See also Human trafficking
Seychelles, 108
Shipping containers
  network ontologies and, 120–121
  regulating criminal economies and, 103–105
Shower Posse, 76
Single Convention on Narcotic Drugs (1961), 157
Small group organization, 73
Smuggling
  counterfeiting and, 87
  criminal organization forms and, 73–74
  networked group organization and, 85
Social factors
  criminal organization and, 5–6, 84
  local geographical contexts and, 113t

production of organized crime and, 13–15
  responding to organized crime and,
    163–164
Social media, 58–60
Sociocultural contexts, 91–92
Sociological perspective, 127–128
Somalia
  criminal organization and, 90–91
  local geographical contexts and, 114
  responding to organized crime and, 166
The Sopranos (1999–2007), 98–99
South Africa, 134
South Asia, 29
South Sudan, 26f
Southeast Asia, 24, 90–91
Soviet Union
  criminal organization and, 83–84
  multiscalar approaches and, 117–118
  overview, 81
  regulating criminal economies and,
    102–103
  See also Russia
Space, 72–73. *See also* Criminal mobilities
Spaces of transit, 124–125. *See also* Criminal
  mobilities; Transit of commodities
Spatial patterns
  global economy and, 100–110
  local geographical contexts, 110–116, 113t
  multiscalar approaches, 117–118
  network ontologies, 119–121
  overview, 18–19, 98–100, 121–122,
    182–183
  regulating criminal economies and, 93–94
  See also Globalization; Migration; Mobility
Stability, 71, 153–154
Standards, 167
State economies, 35–38
State governance
  criminal organization and, 91
  local geographical contexts and, 112,
    113–116, 113t
  responding to organized crime and,
    162–164, 176–177
Strategic Hub on Organized Crime, 67
Strategic location, 113t
Subculture of organized crime, 4–6
Sudan, 26f
Supply chain
  criminal organization forms and, 75
  decriminalization and legalization
    movements and, 171
  responding to organized crime and, 159
Suppression of supply
  drug economies and, 157–158
  responding to organized crime and,
    159–160
Surveys
  organized crime indexes and, 51
  perception and experience surveys, 49–50
Swaziland, 134–135
Sweden, 170

Switzerland, 108
Sydney, 108
Symbolic representation, 84
Synthetic drugs, 28. *See also* Drugs

T

Taboos, 92
Taiwan, 105
Tax havens
  decriminalization and legalization
    movements and, 171
  global economy and, 106–110
Technological advances
  local geographical contexts and, 113t
  responding to organized crime and,
    155–156
Terminology, 9
Terrain, 113t
Territory
  criminal organization and, 76, 80–82,
    89–90
  global economy and, 39
  local geographical contexts and, 112,
    114–115
  regulating criminal economies and, 93
  responding to organized crime and,
    158
Thailand, 24
Theft of the Nation (Cressey, 1969), 6–7
Theories of change, 154, 155f
Threat assessments, 53
Tivoli Gardens, 114–115
Tobacco products, illicit trade in
  decriminalization and legalization
    movements and, 171
  economies of, 22, 29–30
  regulation and, 105
Trade routes, 130–150, 144f, 145t, 149f
Traditions, 113t
Trafficking of commodities
  counterfeiting and, 87
  criminal commodity movements and,
    139–140
  criminal organization and, 75, 85
  networked criminal mobilities and,
    126–127
  networked group organization and, 85
  overview, 4–5, 181–182
  regulating criminal economies and. *See also*
    Drugs
  See also Human trafficking
Transit of commodities, 124–125. *See
  also* Criminal mobilities; Shipping
  containers; Trafficking of commodities
Transnational corporation criminal
  organization form
  criminal commodity movements and,
    138–139, 146–147
  criminal mobilities and, 151
  drug economies and, 157–158
  global economy and, 101

Transnational corporation criminal organization form (*continued*)
  networked criminal mobilities and, 126–130
  networked group organization and, 85–89
  overview, 12, 74, 78
  responding to organized crime and, 155*f*, 162–163, 178
Transnational organized crime (TOC)
  decriminalization and legalization movements and, 171
  network ontologies and, 120
  regulating criminal economies and, 103
  responding to organized crime and, 165
Transparency
  regulating criminal economies and, 105
  responding to organized crime and, 163–164
Trucks, 120–121
True crime autobiographies, 56–58
Trust
  criminal organization and, 79–80
  regulating criminal economies and, 93, 94
Turkey, 31–32

U

U. S. Central Intelligence Agency (CIA), 27
United Kingdom
  criminal economies and, 46–48
  criminal organization and, 82, 84
  cybercrime economies and, 32
  networked criminal mobilities and, 127–128
  networked group organization and, 87
  policymaking and, 67, 67–68
  regulating criminal economies and, 94
  social media and online sources, 58–59
  tax havens and, 108
United National Convention Against Illicit Traffic in Narcotic Drugs and Psychotropic Substances (1988), 157
United Nations Office on Drugs and Crime, 167, 183
United States
  criminal organization and, 81
  cybercrime economies and, 32
  decriminalization and legalization movements and, 170
  drug economies and, 28
  economies of organized crime and, 22
  human trafficking economies and, 30–31
  networked criminal mobilities and, 129–130
  networked group organization and, 87
  regulating criminal economies and, 104–105
  responding to organized crime and, 161
  spatialities of organized crime, 98
  tax havens and, 108
Universal policy regimes, 159–162
Universalism, 88–89
Uruguay, 170
Used car trade, 133–135, 136, 139–140

V

Vans, 120–121
Various noncapitalist economies, 1–3, 2*f*
Venezuela
  organized criminal markets and, 35
  responding to organized crime and, 159
Vietnam
  criminal organization and, 82–83, 89–90
  regulating criminal economies and, 93
Violence
  counterfeiting and, 87
  criminal commodity movements and, 130–132
  local geographical contexts and, 113*t*, 115–116
  regulating criminal economies and, 92–96
  responding to organized crime and, 153–154, 160
Vory v zakone, 83–84
Vulliamy, Ed, 58

W

War on Drugs, 156–157, 161–162, 168–169, 179–180. *See also* Drugs; Policymaking
Warehouses, 120–121
Waste and e-waste recycling, 139–140
Weapons, access to
  local geographical contexts and, 113*t*
  state weakness and, 163
West Africa, 177
Wikileaks, 60
World Drug Report, 147–148

Y

Yakuza, 76
Yugoslav wars of the 1990s, 30

Z

ZeroZeroZero (Saviano), 55–56
Zetas, 94–95

ABOUT THE AUTHOR

Tim Hall, PhD, is Professor of Interdisciplinary Social Studies and Head of the Department of Applied Social Studies at the University of Winchester, United Kingdom. He is a human geographer who was originally interested in urban geography but who has, since 2006, developed research interests in economic geography and contemporary globalization, with a focus on the roles of organized criminal groups. He is author or editor of a number of articles, book chapters, and books on organized crime and globalization, urban geography, and pedagogy. Dr. Hall is a recipient of the Award for Promoting Excellence in Teaching and Learning from the *Journal of Geography in Higher Education* and the Excellence in Leading Geography Award from the Geographical Association, the latter for his article "Where the Money Is: The Geographies of Organized Crime."